KT-484-821

100 ANATOMY & STRETCHING EXERCISES FOR Cycling

ANATOMY & 100 STRETCHING EXERCISES FOR Cycling

First edition for the United States, its territories and dependencies, and Canada published in 2016 by Barron's Educational Series, Inc.

English translation by Eric A. Bye, M.A.

Original Spanish title: *Anatomía & 100 Estiramientos Esenciales Para Cycling*

Published by Parramón Paidotribo, S.L., Badalona, Spain

All inquiries should be addressed to:
Barron's Educational Series, Inc.
250 Wireless Boulevard
Hauppauge, NY 11788
www.barronseduc.com

ISBN: 978-1-4380-0858-5

Library of Congress Control No.: 2015947874

Printed in Spain
9 8 7 6 5 4 3 2

Paidotribo Publishing is grateful to Dr. Victor Götzens García, professor of human anatomy with the medical faculty of the University of Barcelona, for his help with the scientific review of the text. Also, the publisher appreciates the contributions of Anna Flaquer Porti, Alin Tani Catarig, Cristina Caballero Nogales, and Eduard Massó Grabulosa, the models for the sports photos.

Project director: Paidotribo Publishing
Editorial Director: Maria Fernanda Canal
Scientific Review: Professor Victor Götzens García
Text: Guillermo Seijas
Proofreading: Ana Lorenzo, Roser Pérez
Graphic Design: Toni Inglès
Illustrations: Myriam Ferrón
Photography: Nos i Soto
Layout: Estudi Toni Inglès

Preface

Whether you are a devoted or casual cyclist, you certainly know the pleasure and the convenience that a bicycle offers. Bicycles may also bring up memories of your childhood, vacations, and weekends with friends.

Do you remember your first bicycle? Of course you do. Maybe it was an old bicycle handed down from a relative—a heavy steel bicycle that was uncomfortable and difficult to ride, but still you have nothing but fond memories. Maybe that bicycle was passed on to a younger brother and, later, on to your cousins, and, today, it still remains in the family, like new, after years of livening up the summers of several generations.

Now you have this book in your hands, and this is a sure sign that your enthusiasm for riding remains intact and that you yearn to know more, with that desire for continual improvement that motivates all athletes.

When cycling turns into a hobby, it becomes essential to learn more about exercise. Just as cycling has improved over the years, so has the knowledge of mechanics, training, specific ailments, biomechanics, and physiology.

Knowing about exercises related to cycling is necessary for you to achieve optimal performance, while avoiding ailments and injuries common to cyclists. It doesn't matter whether you want to reach an elite level, extend your athletic life to the maximum, or have fun while minimizing the risks. The information that this book provides will help you, and, even though it delves into some complex issues, it is written for easy comprehension, regardless of how much you know about cycling at this point in your career.

In these pages, you will find techniques for setting up your bicycle to match your body shape, information to help you understand possible causes of some of the most common ailments in cycling, and exercises that will help you to avoid them. You will also learn to distinguish between different levels of pedaling momentum, to distinguish among the main muscles involved in cycling, from the most obvious ones to the ones that we don't notice (even though the latter play an essential role), and to undertake the best stretching exercises to keep these muscles in perfect condition and to avoid strain, especially for those who cycle regularly.

In these pages, you will find a selection of exercises and tips for improving your athletic experience and your cycling, based on the premise that health and athletic performance go hand in hand.

Contents

How to Use This Book

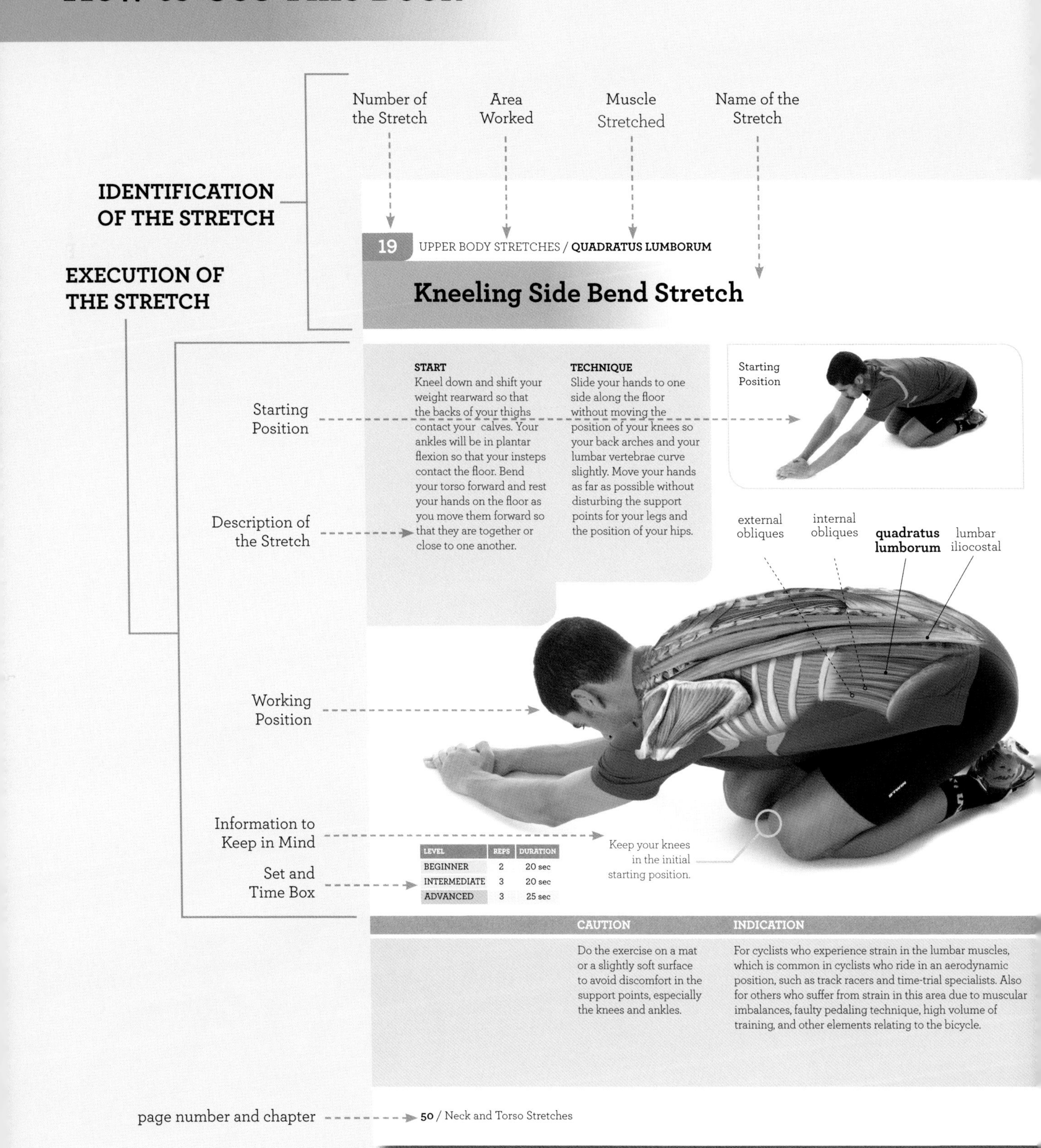

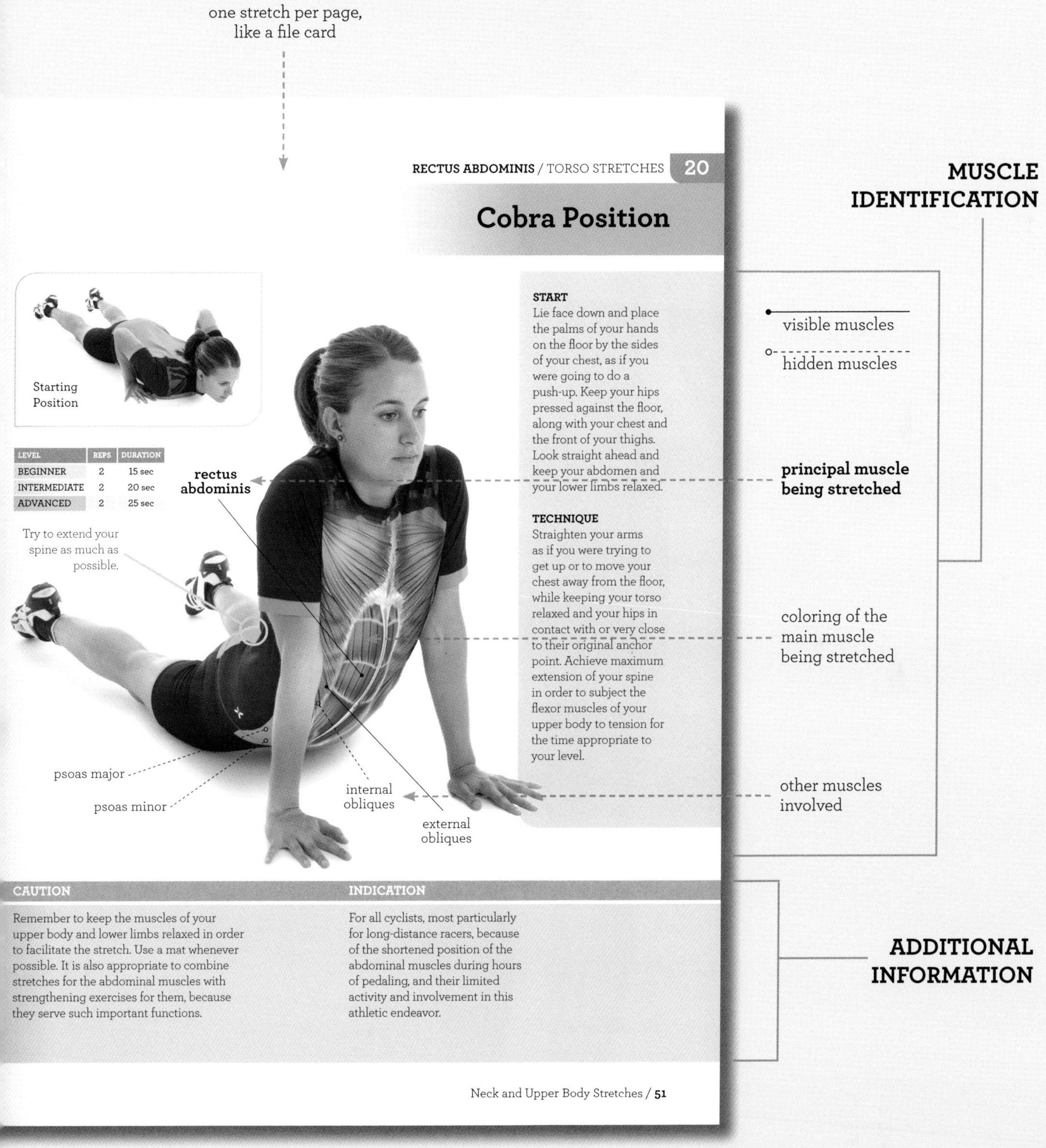

one stretch per page, like a file card

RECTUS ABDOMINIS / TORSO STRETCHES 20

Cobra Position

Starting Position

LEVEL	REPS	DURATION
BEGINNER	2	15 sec
INTERMEDIATE	2	20 sec
ADVANCED	2	25 sec

Try to extend your spine as much as possible.

rectus abdominis

psoas major

psoas minor

internal obliques

external obliques

START
Lie face down and place the palms of your hands on the floor by the sides of your chest, as if you were going to do a push-up. Keep your hips pressed against the floor, along with your chest and the front of your thighs. Look straight ahead and keep your abdomen and your lower limbs relaxed.

TECHNIQUE
Straighten your arms as if you were trying to get up or to move your chest away from the floor, while keeping your torso relaxed and your hips in contact with or very close to their original anchor point. Achieve maximum extension of your spine in order to subject the flexor muscles of your upper body to tension for the time appropriate to your level.

CAUTION
Remember to keep the muscles of your upper body and lower limbs relaxed in order to facilitate the stretch. Use a mat whenever possible. It is also appropriate to combine stretches for the abdominal muscles with strengthening exercises for them, because they serve such important functions.

INDICATION
For all cyclists, most particularly for long-distance racers, because of the shortened position of the abdominal muscles during hours of pedaling, and their limited activity and involvement in this athletic endeavor.

Neck and Upper Body Stretches / 51

MUSCLE IDENTIFICATION

visible muscles

hidden muscles

principal muscle being stretched

coloring of the main muscle being stretched

other muscles involved

ADDITIONAL INFORMATION

Anatomical Atlas
The Locations of the Muscles

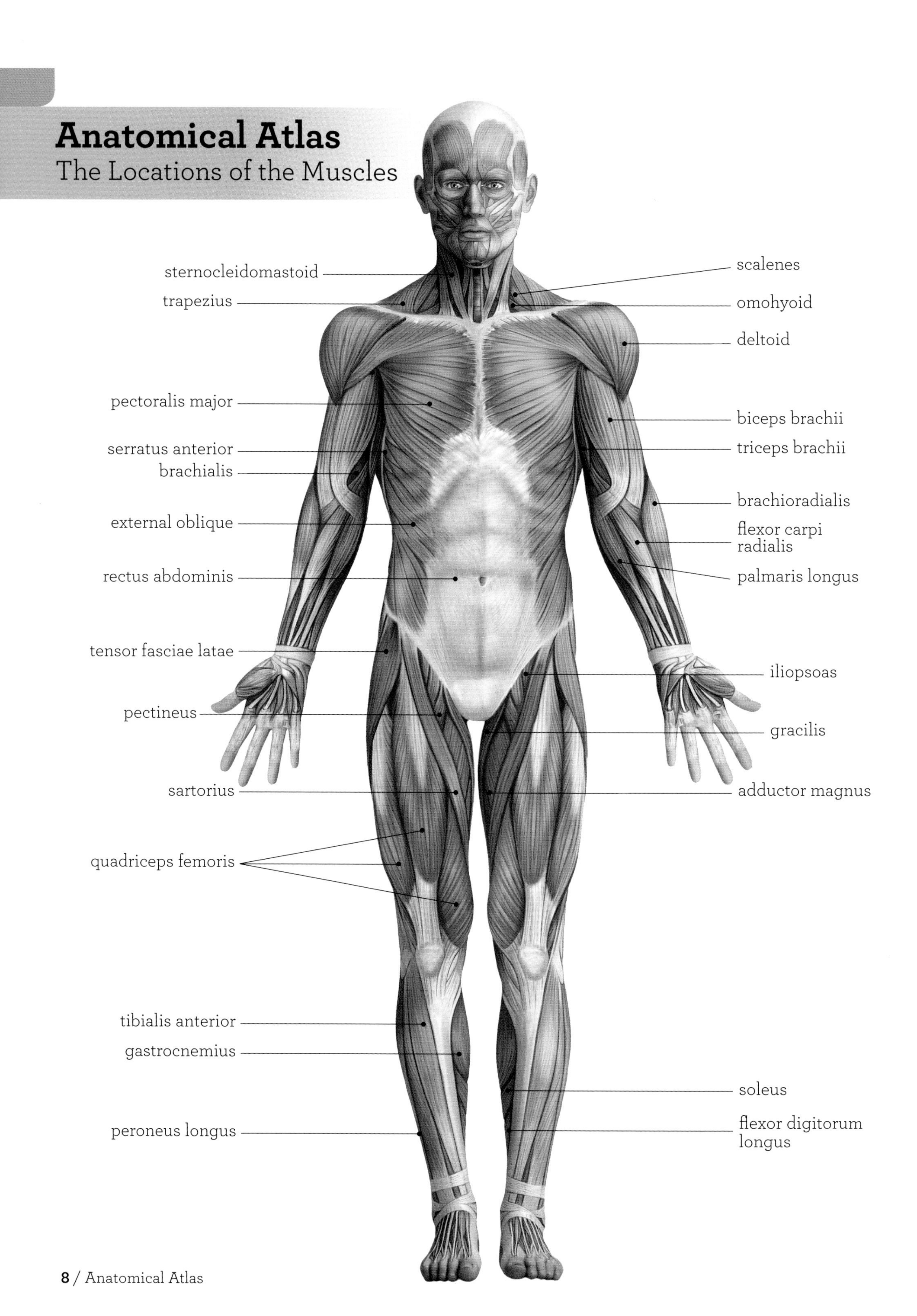

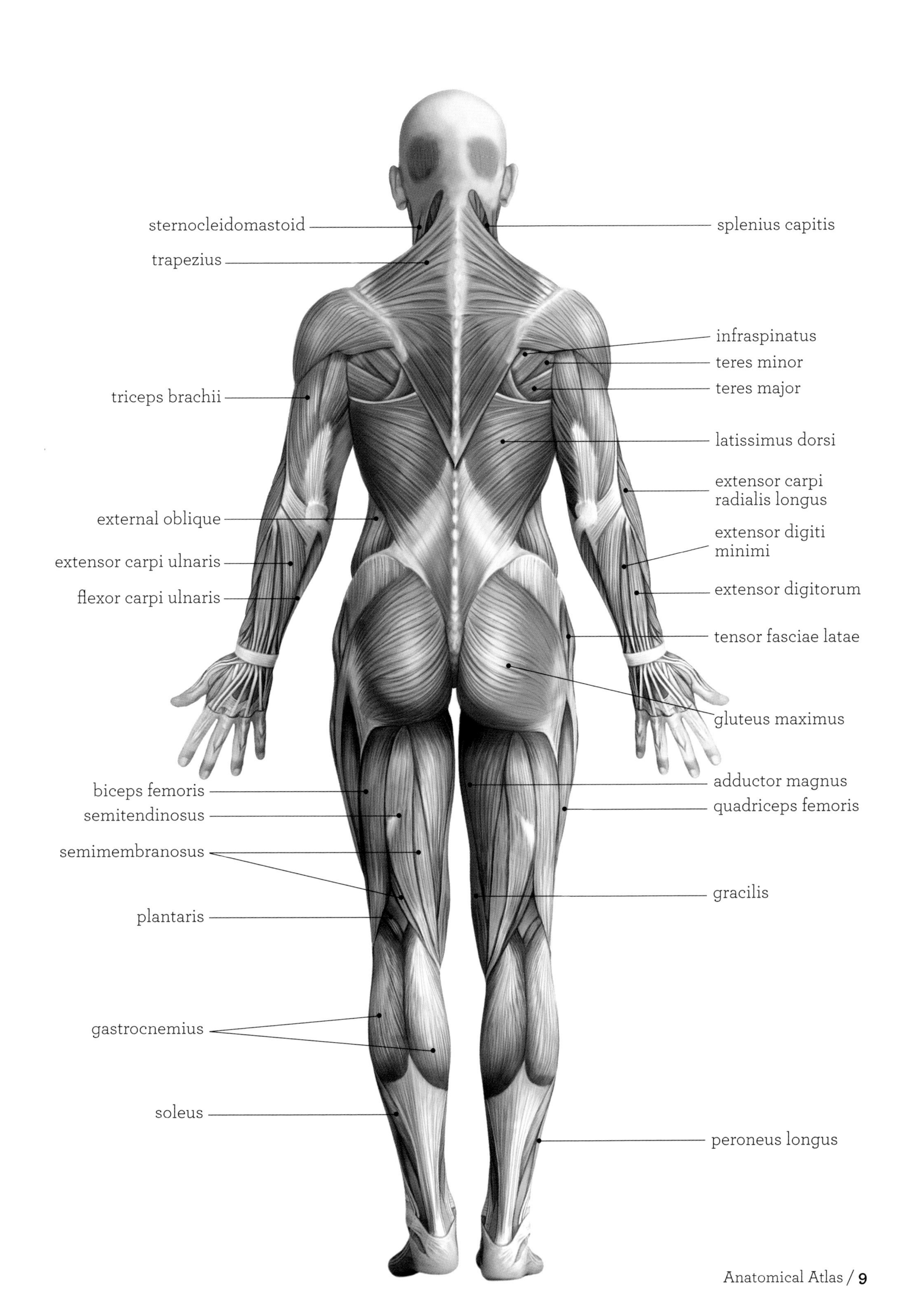
sternocleidomastoid
trapezius
triceps brachii
external oblique
extensor carpi ulnaris
flexor carpi ulnaris
biceps femoris
semitendinosus
semimembranosus
plantaris
gastrocnemius
soleus
splenius capitis
infraspinatus
teres minor
teres major
latissimus dorsi
extensor carpi
radialis longus
extensor digiti
minimi
extensor digitorum
tensor fasciae latae
gluteus maximus
adductor magnus
quadriceps femoris
gracilis
peroneus longus

Planes of Movement

Before we start, it is appropriate to explain a series of terms that refer to body movements, and which appear in recurring fashion throughout the book. If you do not know the basic nomenclature of the movements, it will be difficult to understand the detailed descriptions of the exercises. Some of these terms, such as *bending* and *extending* are in common usage, but others, such as *inversion, eversion, adduction,* and *supination,* are often used in narrower circles, so it may be useful to review their meanings.

The first thing we need to know is that body movements take place in three different planes: the frontal, the sagittal, and the transverse planes. There is a certain group of movements that correspond to each plane, as we will see below. We can begin to understand them with the basic anatomical position in the picture.

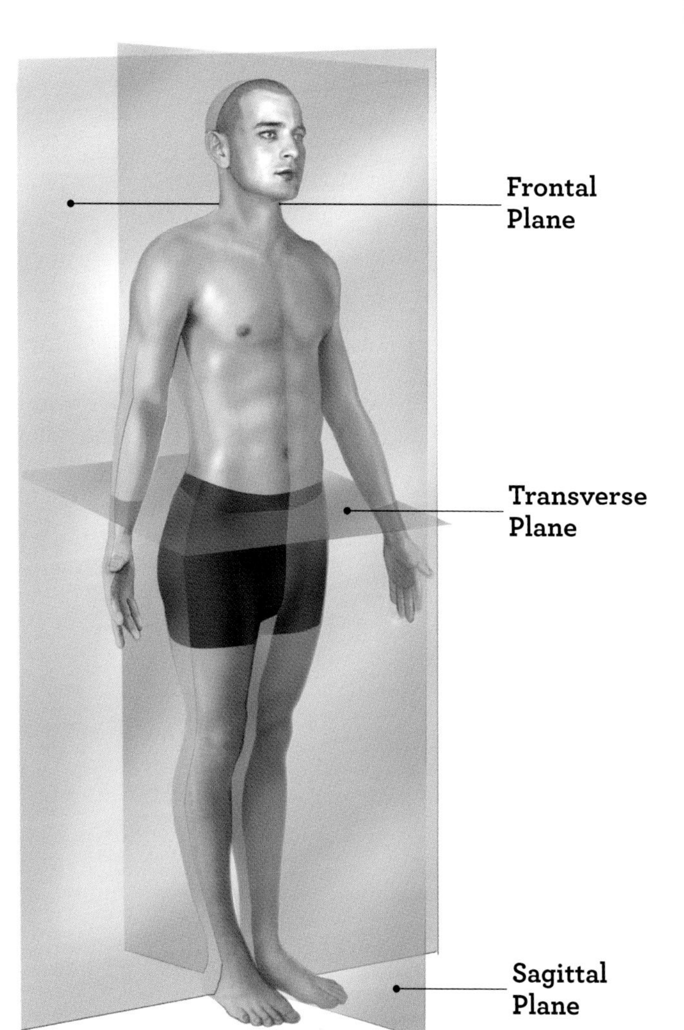

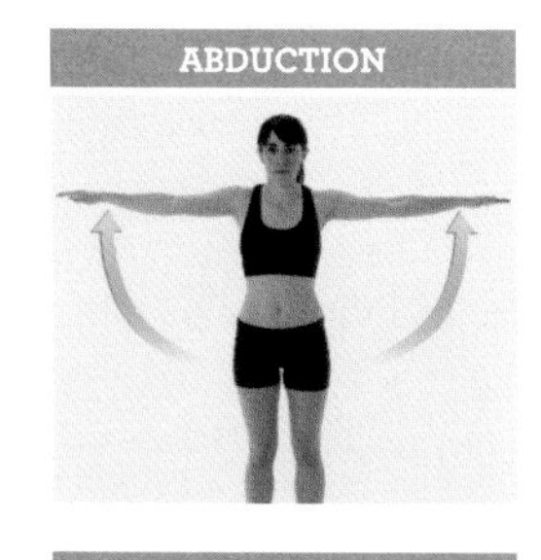

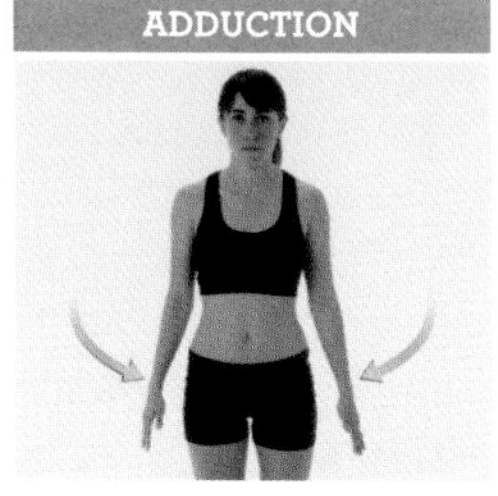

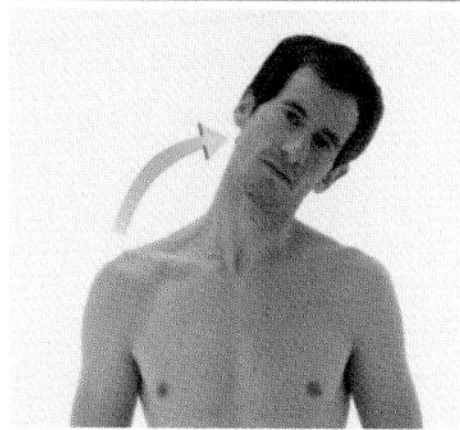

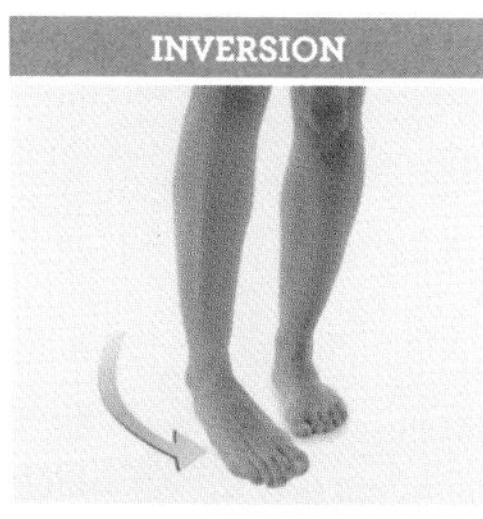

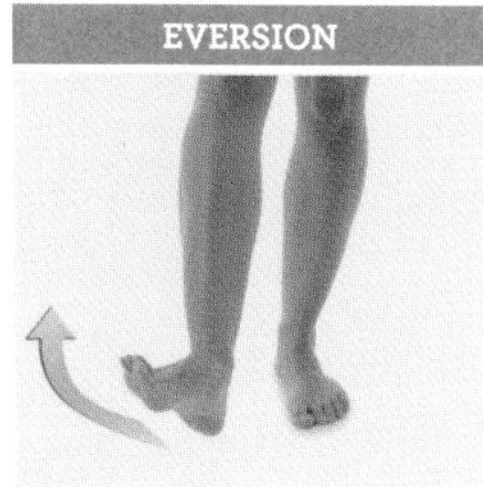

FRONTAL PLANE

This divides the body into ventral and dorsal planes—in other words, front and back. The chest and the stomach are in the ventral plane, and the back of the neck, the back, and the buttocks are in the dorsal plane. The movements in the frontal plane are as follows:

Abduction: a movement in which you move a limb away from the central axis of the body. This is easily seen from the front or the rear, because the variation in the outline of the body is easy to see from this perspective. When you hold your arm straight out to the side, you perform abduction of the shoulder.

Adduction: a movement in which you bring a limb toward the central axis of the body—in other words, the movement opposite abduction. If you stand with your arm straight out to the side and you lower it so that is close to your body, you perform adduction of the shoulder.

Lateral Inclination: a movement in which you tilt you head, you neck, or your upper body to the side. If you fall asleep in a sitting position, generally your head and neck end up tipping to one side through lateral inclination.

Inversion: although this movement does not belong solely to the frontal plane, this is where it is most common. Foot inversion occurs when the tip and the sole are placed toward the inside while performing plantar flexion.

Eversion: a turning movement toward the outside—for example, the tip and the sole of the foot face outward while performing dorsal flexion.

FLEXION

EXTENSION

ANTEPULSION

RETROPULSION

DORSAL FLEXION

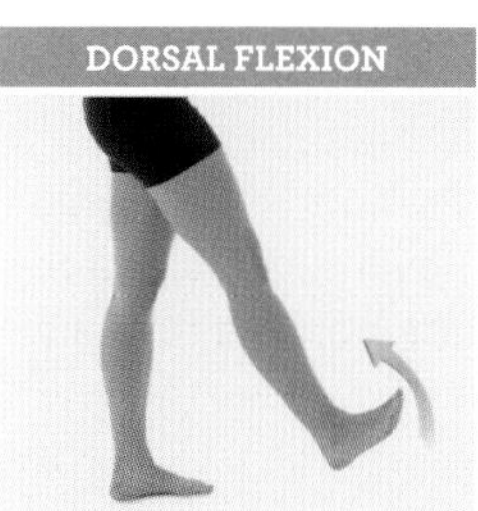

PLANTAR FLEXION

SAGITTAL PLANE

This divides the body into two halves: right and left. Movements in this plane are best perceived by looking at the individual's profile from one side. The following movements are highlighted in this plane:

Flexion: a movement in which you move one part of the body ahead with respect to the central axis. There are exceptions to this definition, such as knee flexion and plantar flexion of the ankle.

Extension: a movement in which you move a part of the body rearward with respect to the central axis, or you align it with the axis. For example, if you stand and look up at the sky, you have to perform an extension of the cervical vertebrae. Once again, the knee is an exception.

Antepulsion: This is equivalent to flexion, but it is applicable solely to the movement of the shoulder.

Retropulsion: This is equivalent to extension, but it is applicable solely to the movement of the shoulder.

Dorsal Flexion: a flexion movement that is applicable solely to the ankle joint.

Plantar Flexion: the term used to designate the ankle movement equivalent to extension.

EXTERNAL ROTATION

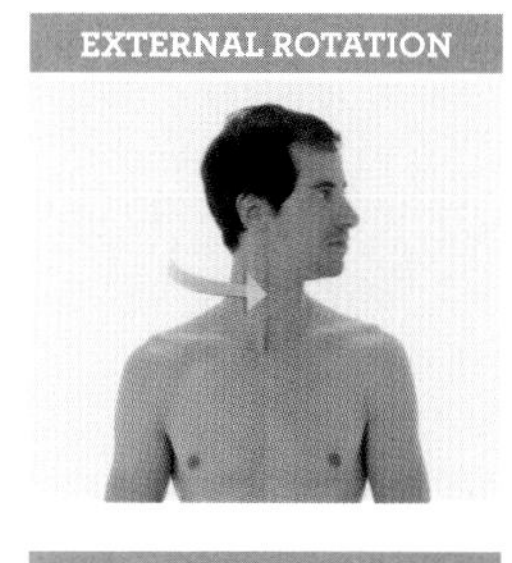

INTERNAL ROTATION

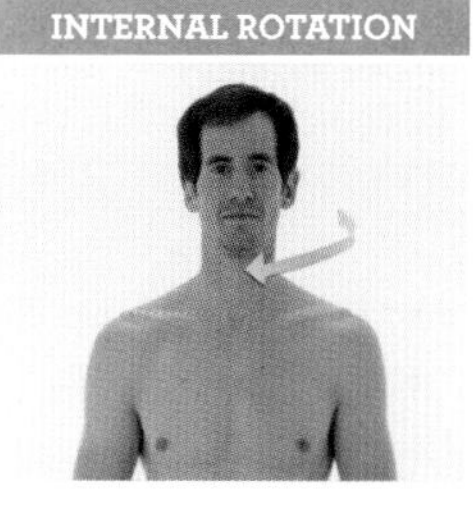

PRONATION

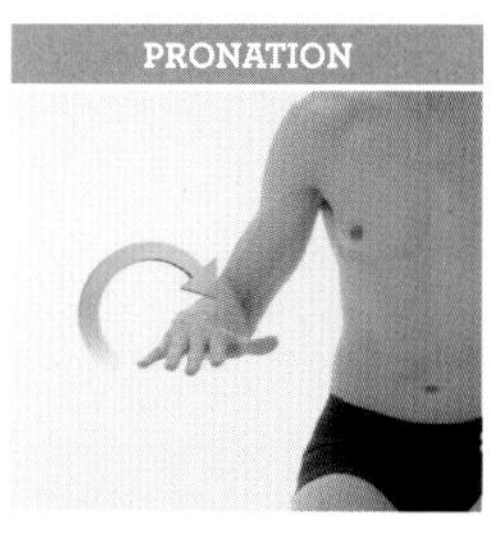

SUPINATION

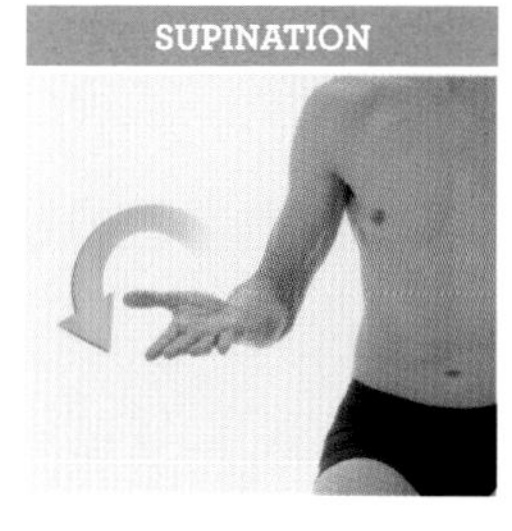

TRANSVERSE PLANE

This divides the body into an upper and a lower part. Movements in this plane are easily seen from any vantage point, although somewhat better from above or below the individual. They are as follows:

External Rotation: a movement in which you turn a body part toward the outside and along its axis. If you are seated at a table and the person next to you speaks to you, you perform an external rotation of the neck to look at him as he or she speaks.

Internal Rotation: a movement opposite the previous one, because it involves turning a part of the body toward the inside along its axis. When you complete the conversation with the person sitting next to you, you perform an internal rotation of the neck to return your gaze toward the front.

Pronation: a rotational movement of the forearm, in which you place the back of the hand upward and the palm facing downward. When you use a knife or a fork to manipulate the food on a plate, your hands are in pronation.

Supination: a movement opposite the preceding one, which involves rotating the forearm such that you place the palms facing upward. For example, if someone gives you a handful of sunflower seeds, you place your hands palms-up, in supination, similar to a bowl, so you don't drop them.

A Brief History of the Bicycle

People have always sought new ways to get around, whether on land, on water, or in the air. Their eagerness to invent contraptions for traveling about is merely the reflection of a necessity: the need for freedom, a longing for autonomy, independence, plus the excitement of speed.

Even though time and technological innovation have enabled us to use highly sophisticated tools to travel from one point to another, we have always felt a special satisfaction in using tools that constitute extensions of our bodies, in which the main driving force is our bodies themselves. Starting in childhood, we have experienced the enjoyment of gliding on a skateboard, on skates, a scooter, or a bicycle: controlling a reliable tool whose simplicity is also its strength. Light, small, and maneuverable tools provide an extraordinary number of possibilities, such as forward motion, jumping, skidding, turning, climbing mountains, or going down very steep hills in a matter of seconds.

The bicycle is merely one of these tools that have facilitated people's travel, fun, and autonomy, and various individuals have attempted to assign this worthy invention to their respective countries, even falling victim to error, whether intentionally or otherwise, as happened with the bicycle supposedly designed by Leonardo da Vinci, the sketch of which "appeared" inserted in his *Codex Atlanticus*, or with the attribution of the *célérifère* to Count Mede de Slivrac. Even though its origin is unclear, this *célérifère* certainly existed, because there are records of its use in the early years of the 19th century, and even though it cannot be considered the first bicycle, because it had no steering mechanism and movement was imparted by pushing against the ground with the feet, as on a scooter, it nevertheless has some similar characteristics, such as wheels in alignment with one another, and a saddle.

However, evidence has been found of the existence of the first contraptions with two aligned wheels as early as the Egypt of the pharaohs, in ancient China, and in Indian and pre-Columbian cultures. It's certain that people and bicycles share a long history, and that this invention has been a recurring feature in very different times and geographical locations. The appearance of the modern bicycle, in short, was probably just a matter of time.

Just the same, it seems certain that the origin of the modern bicycle was in Europe. At the start of the 19th century, Karl Freiherr von Drais invented a device that was called the *draisine*, which contributed a new feature to what had previously existed: steering of the front wheel, which was accomplished by means of its connection to a rudimentary handlebar. The *draisine* enjoyed popularity for some years, because it involved a noteworthy innovation, even though it was hard to operate and offered few advantages for traveling. It was even copied on many occasions.

Later, around 1840, the first bicycle design including pedals appeared; it is attributed to the Scotsman Kirkpatrick MacMillan. Propulsion for this bicycle was produced by a system of rods that transmitted the movement to the rear wheel.

More than 20 years after, Frenchman Pierre Michaux began selling a bicycle with pedals that were installed on the hub of the front wheel, which was the drive wheel. Still, it is not clear whose idea this was, because some historians attribute the design to Pierre Lallement, a young man who

Draisine *by Karl Freiherr von Drais.*

may have collaborated with Michaux for a brief time and who subsequently moved to the United States, where he recorded the first bicycle patent on this continent.

As early as the 1870s, James Starley came much closer to the modern bicycle. This Englishman, who began his career on the family farm, soon made a name for himself for his skill in making and repairing machines, which gained him a job at Newton Wilson & Co., a company that focused on manufacturing sewing machines. Soon thereafter he founded his own company and began designing velocipedes. His first designs were made of iron and used solid tires with a significant difference in size between front and rear; they enjoyed great acceptance, but he kept working tirelessly to come up with better bicycles.

Velocipede by James Starley.

Later on, the sprung saddle was introduced, along with frames made from hollow steel tubing, the chain drive, an equalization of the size of the front and rear wheels, and many other advances that led to what we now know as a bicycle.

Bicycles have continued evolving from yesterday's iron beasts weighing more than 88 pounds (40 kg) to today's modern carbon-fiber bicycles that weigh less than 22 pounds (10 kg), and, what's more, they have also become specialized and adapted to different sports and types of terrain.

Road bicycling is now one of the most common forms of cycling, and it may also be the oldest one, for the great road cycling stage races are over 100 years old. From the first Tour de France in 1903 through the most recent race, we have witnessed a history beset with great myths and epic achievements that will remain in historical memory. The arduous nature of stage races—such as the Tour de France, the Giro d'Italia, and the Vuelta a España—have turned competitive long-endurance cycling into a particularly epic sport that is admired by aficionados and the general public.

Other bicycle sports are no less important, such as track racing, mountain biking, and cyclocross (BMX), with their different events, and forms of bicycling more closely connected to recreation, bicycle touring, and the use of leisure bikes for urban travel. Each of these sports requires bicycles with different characteristics, which also affect their sizes, weights, tire treads, gears, braking systems, shock absorption, and construction materials, and, of course, the position of the cyclist on the bicycle.

The bicycle, to sum up, continues to be a widely accepted means of travel, because of its simplicity, its affordable acquisition and maintenance costs, and the enjoyment it provides. Additionally, with every year that goes by, we are more conscious of the fact that we need to start moving about in a more sustainable way that respects the environment. The authorities of many cities understand the contribution that bicycle use can make to sustainability and the proper functioning of medium-sized and large urban centers, and there are more and more cities that provide dedicated bicycle lanes and bike rental services for use by the general public.

Bicycle from the early 20th century.

Adjustments

Bicycles appear to have nothing but advantages, but using them requires a certain amount of knowledge to choose them, to get the right size, materials, and characteristics, and to adjust and use their components. This becomes particularly clear when you change from sporadic or recreational use to competitive use of the bicycle, so knowledge can help you avoid problems that result from this change.

Size. The size of the bicycle is a significant determinant in adjusting it properly for athletic use. If you choose a very large bicycle, you may not be able to lower the saddle or the handlebar far enough to pedal comfortably, and, if you choose a very small one, you will be able to adjust the height of the saddle and the handlebar, but pedaling will remain uncomfortable, because of the improper length of the cranks. To avoid these problems, you can apply a very simple formula. First of all you measure the height in centimeters between your crotch and the ground, which we will call PG (it's a good idea to take off your shoes and separate your feet slightly). We multiply this measurement by 0.56 in the case of mountain bikes and by 0.65 for racing bikes, and the resulting number is the size of the bicycle you need. If the size is not specified on the frame, you can figure it out by measuring the distance on the seat tube between the crank axle to the center of the seat post clamp. You must remember that the number obtained using the formula to determine your bicycle size may not coincide precisely with the available sizes, so you have to choose the closest size. Many times mountain bike sizes are indicated in inches, so to determine the size in inches, you have to multiply the PG by 0.22. Another method for determining the size that is less reliable but simpler is to straddle the bicycle: the distance between your crotch and the frame should be about 1 to 2 inches (2.5–5 cm) for road bikes and 2.5 to 3 inches (6–7.5 cm) for mountain bikes. Clearly, this is a rough way of calculating approximate size, given the variety of current frame designs and the sloping top tubes of many models. The more the top tube slopes, the less reliable this way of calculating the size will be.

Approximate distance between the top tube and the crotch, or perineum.

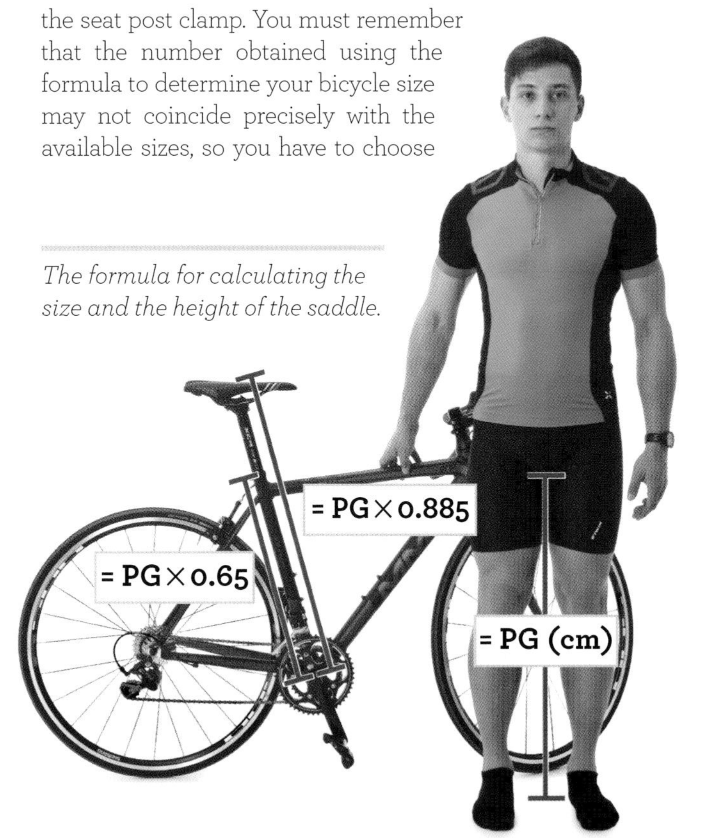

The formula for calculating the size and the height of the saddle.

SADDLE

Saddle Height. This is an important item to consider, because cycling with the saddle at the wrong height could cause pain, mainly in the knees, but also in other places. The lumbar vertebrae and the ankles may also be affected in some cases. The saddle height can be calculated in three ways that are equally reliable, even though none of them is 100 percent accurate, given that every cyclist has a different pedaling style. You can play with the saddle height by raising or lowering it very slightly with respect to the measurement.

The first way to figure out the correct saddle height involves determining the PG once more. This corresponds to the dis-

With the cranks aligned with the seat tube, the bend in the knee should be around 150°.

If you place your heel on the pedal, the knee should be nearly straight when the pedal is at the bottom of its stroke.

tance in centimeters from the perineum to the ground, with no shoes on. The number produced by this measurement is multiplied by 0.885, and the result is the required distance from the crank shaft to the upper surface of the saddle, keeping the measuring tape parallel to the seat tube and the seat post.

A slightly more complicated way to calculate the proper saddle height involves measuring the angle between the femur and the tibia of the lower leg when the cranks are aligned with the seat tube. This angle should be around 150°.

A slightly simpler way of adjusting the saddle, which requires no measuring tape or calculator, involves getting onto the bicycle and resting your heels on the pedals. When the pedal is at the bottom of its stroke, your knee should be almost completely straight, but without being locked straight.

A very high saddle may cause the hips to rock from side to side and detract from pedaling efficiency. It may also produce lumbar discomfort and knee pain. Pain in the medial, lateral, and posterior areas of the knee may come from ailments, such as tendonitis of the pes anserinus ("goose foot"), a syndrome involving the iliotibial tract, and tendonitis in the biceps femoris, respectively. All of these may be caused by a saddle that is too high or too far to the rear. However, you also must avoid a saddle position that is too low or too far forward, because the effects are equally negative, mainly on the knee. A saddle too far forward could cause tendonitis of the patella and the quadriceps, and chondromalacia patellae. All of these show up as pain in the front part of the knee.

Saddle Position. First of all, the saddle must be horizontal. It is very easy to make the saddle horizontal: all you have to do is place a spirit level on top of it to span from front to rear (the ground under the bicycle, of course, must be level, too). To determine whether the saddle needs to be moved forward or rearward, you get onto the bicycle and move the cranks to the horizontal position. The front of the forward knee should be in line with the axis of the pedal. To test this, you can use a plumb line suspended from the front of the knee. To make the plumb line sufficiently reliable, you simply need to use a string with a fairly heavy item (e.g., an eraser or pen) tied to one end.

Keep in mind that a saddle that is positioned too far to the front or the rear can cause the same discomfort, especially in the knee, as a position that is too high or too low.

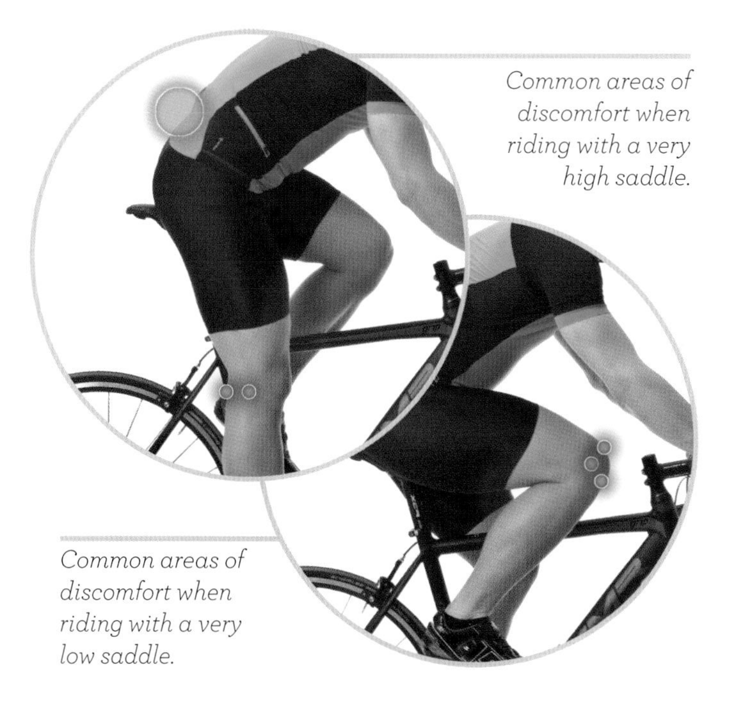

Common areas of discomfort when riding with a very high saddle.

Common areas of discomfort when riding with a very low saddle.

When the cranks are horizontal, the forward knee should be in line with the axis of the pedal.

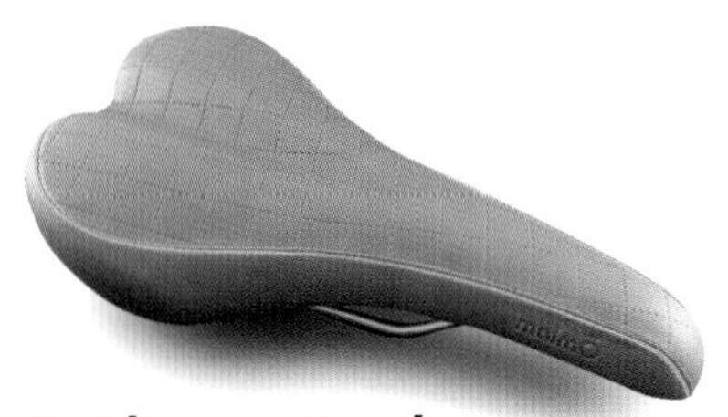

A wide recreational saddle: *Wider and softer.*

Duopower saddle: *It's shorter and has a central channel. These two characteristics reduce pressure on the perineum. The flat model (left): for cyclists who do not have optimal flexibility in the ischiotibial muscles. The curved model (right): for cyclists with good flexibility in the ischiotibials.*

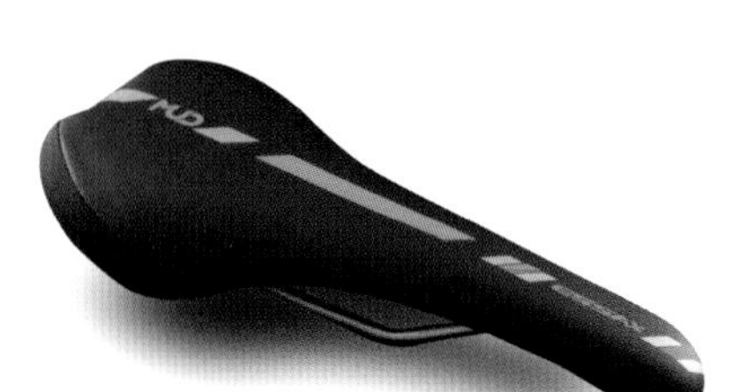

Flat, narrow, and light saddle: *for cyclists who contact the saddle at the rear and have no problems with sensitivity from pressure on the perineum.*

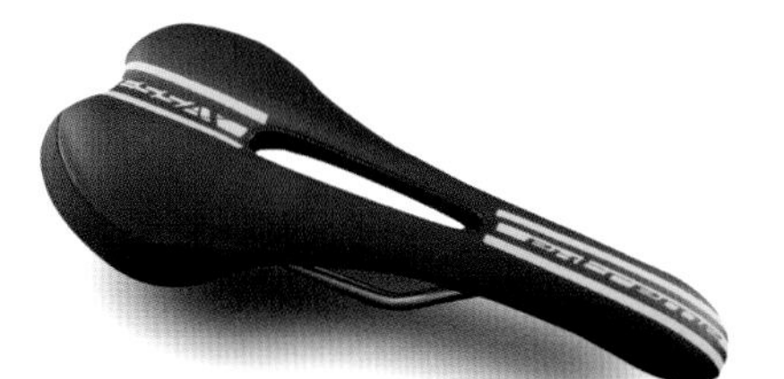

Saddle with central opening: *for relieving pressure on the perineum.*

Shark saddle: *improves the cyclist's perception of the position and contact points on it.*

Saddle Types. The way you use your bicycle and the type of bicycle you have greatly influence the saddle you need. The saddle is one of the points on which you rest your weight, and, over time, using the wrong one can lead to circulatory and nerve problems due to the pressure exerted on the perineum and the pudendal nerve, which can cause lack of sensitivity or numbness in the genitals, especially in men. For these reasons, choosing the right saddle is crucial, especially for people who ride a lot. It's appropriate to remember that, as long as you stay within the characteristics appropriate to a saddle for your sport, comfort must take precedence over all other considerations, such as weight and appearance.

First of all, you need to deal with your position on the bicycle. If you are to adopt an erect position, the support point will be the lower and rear part of the pelvis, with the support on the ischium, and, to a lesser degree, the sacrum. In this position, the pressure on the perineum is relatively minor. As your position on the bicycle requires leaning the upper body forward, the support point changes. A support point farther forward, as on a racing bicycle, produces more pressure on the perineum, and, in order to alleviate it, you use saddles with more pronounced curvature that are higher at the rear. Gel cushions and saddles that have a central opening, or openings, that may reduce the pressure on the perineum are also appropriate for road and mounting biking, although to a lesser degree.

The type of support on the saddle is also determined by the flexibility of the cyclist's ischiotibial muscles. Good flexibility in the ischiotibials allows the cyclist to lean the upper body forward by bending at the hips, so the support point is farther forward, and saddles with more curve are called for. Still, cyclists with little flexibility in the ischiotibials will have less ability to bend the hips, so they will achieve an aerodynamic position by bending the upper body and using a contact point farther to the rear on the saddle, as if they were in an upright position. Thus, a flat saddle will be a better choice for their characteristics. Although this last option involving bending the upper body is the least appropriate one due to

The contact point on the saddle varies by bicycle and sport.

the back problems it may lead to, if you are among the cyclists who don't have much flexibility, you have to abide by your circumstances and choose a flat saddle, at least until you improve your flexibility.

Finally, the separation between the cyclist's ischium bones determines the width of the saddle needed. With a greater separation between them, a wider saddle should be chosen, and, with a lesser separation, a narrower saddle. Generally speaking, and solely as a point of reference, women usually have a wider separation between the ischia than men, so they tend to choose slightly wider saddles.

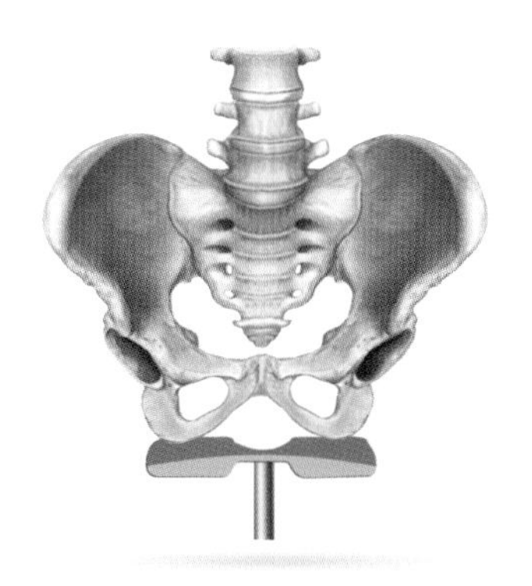

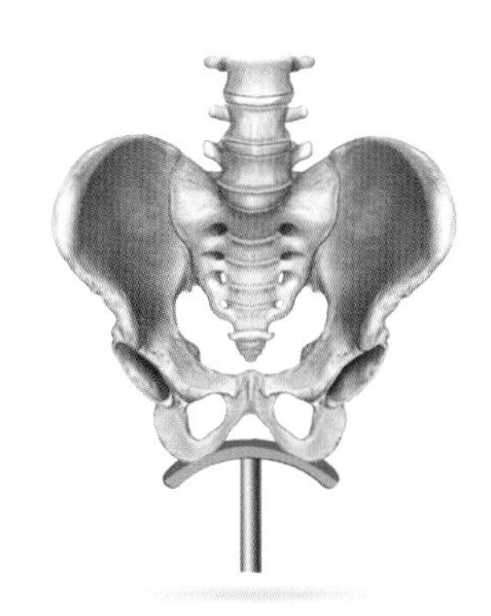

The width of the handlebar should be the same as the rider's shoulder width.

HANDLEBAR

Handlebar Height. Handlebar height is largely determined by the weight distribution on the bicycle and the cyclist's position. A handlebar that is too low involves excess forward lean in the upper body, which can produce excess pressure on the hands, resulting in arm fatigue, and even numbness in the hands, due to compression of the median and carpal nerves. At the same time, this excessive lean may involve strain in the spinal erectors, especially the cervical ones, due to the hyperextension of the neck required for keeping the cyclist's gaze directed forward. Excessive lean can also cause problems in the spine and its structures, including the spinal erectors: in this case, the ones along the lumbar and thoracic sections of the spine, due to the constant bend to which these areas are subjected. Finally, the gluteus muscles, with the greater bend in the hips, may experience strain and reduced performance. In spite of this, it is very common to see cyclists with handlebars that are too low and saddles that are too high, in an attempt to imitate and even exaggerate the position of elite road racers.

A handlebar that is too high will reduce a cyclist's aerodynamics, which will necessarily hurt his or her times, especially in road and track riding, which involve greater lean over the handlebars. It's easy to tell by eye that not all handlebars should be at the same height. When you see someone on a recreational bicycle, where comfort is the primary consideration, the handlebar is higher, and most of the support comes from the saddle. This is why the saddles on these bicycles usually are wider, cushioned, and much more comfortable. Mountain bikes commonly involve an intermediate position that allows very good control of the front wheel and steering over irregular terrain. Finally, road bikes have lower handlebars with respect to the saddle in order to create greater aerodynamics in motion, because economy of energy output is paramount, and riding the roads does not involve the abrupt maneuvers that cross-country does. In order to determine the height of the handlebar, you use the height of the saddle as a reference point. Generally speaking, on mountain and road bikes, the handlebar is slightly lower than the saddle. The difference in height between the saddle and the handlebar is referred to as *handlebar drop*. There is no exact formula for calculating the handlebar drop with respect to the saddle, so it is determined using a table of general reference points based on the saddle height.

MOUNTAIN BIKE		ROAD BIKE	
SADDLE HEIGHT	HANDLEBAR DROP WITH RESPECT TO SADDLE	SADDLE HEIGHT	HANDLEBAR DROP WITH RESPECT TO SADDLE
65 to 68 cm	1 cm	65 to 68 cm	5 to 6 cm
69 to 72 cm	1 to 2 cm	69 to 72 cm	6 to 7 cm
73 to 76 cm	2 to 3 cm	73 to 76 cm	7 to 8 cm
77 to 79 cm	3 to 4 cm	77 to 79 cm	8 to 9 cm
80 to 82 cm	4 to 5 cm	80 to 82 cm	9 to 10 cm

Handlebar width. From the outset, you need to know that a wider handlebar allows for better control of the bicycle, plus greater aerodynamics. In any case, the recommended handlebar width should be about the distance between the ends of the cyclist's clavicles, to produce a riding position that is comfortable and aerodynamic.

HANDLEBAR STEM

The length of the handlebar stem is an important factor in both comfort and aerodynamics. A shorter handlebar stem provides greater comfort and control over the bicycle while riding, and a longer handlebar stem sacrifices comfort for a slight improvement in aerodynamics. Nevertheless, the choice of a handlebar stem is a very personal option determined by the type of sport you practice. A downhill racer will choose a very different handlebar stem from what a road racer would choose, even though their anthropometric values are similar.

CRANK LENGTH

As with other bicycle components, when you buy a bicycle, the cranks are already in place. Generally, they are keyed to the size of the bicycle, so they are probably the right ones for the cyclist, as long as the size was selected according to the parameters discussed earlier. In any case, you can use the following chart, which is based on the distance from the perineum to the ground.

The crank length is keyed to the size of the frame, so, except in rare cases, it should not be changed for reasons of improper length.

DISTANCE FROM THE PERINEUM TO THE GROUND	CRANK LENGTH
< 73,5 cm	16.5 cm
73.5 to 81.5 cm	17 cm
81.6 to 86.5 cm	17.25 cm
> 86.5 cm	17.5 cm

ACCESSORIES

In cycling, the accessories vary tremendously based on the sport, but there are certain accessories that are always essential, because they are directly related to the cyclist's safety.

Helmet. First of all, you must always wear an approved helmet adapted to your size and sport. In addition, it must be adjusted properly. Like all equipment in cycling—and sports in general—the helmet has a limited lifespan beyond which the materials from which it is made change characteristics. As a result, it must be replaced from time to time, even though it has not been subjected to any impact and the outward appearance is still good. You also need to keep in mind that the construction materials are significant determinants of the helmet's durability. One good example of this is that some steel or aluminum bike frames are guaranteed for life, but the best carbon-fiber frames rarely have a guarantee for more than five years. This means that it's a good idea to get some advice from a technician about the type of helmet you should get based on your needs and your financial means.

Sunglasses may seem to be a totally incidental item, but they are highly recommended, especially if you intend to ride a fair amount or a lot. Good athletic sunglasses not only provide comfort during the ride and keep you from becoming dazzled by the sun. but they are also a protective feature, because they shield your eyes from anything that might be in the air or bounce up from the road surface. Insects and seeds often get into cyclists' eyes, and, in the best cases, they sting and interfere with vision. In the worst cases, they can damage the eyeball and cause a crash.

This means that choosing glasses is important, so you need to select sport glasses made from tough plastic materials that do not break on impact, as could happen with glass lenses. If a pebble launched by a motor vehicle can break a windshield, you can imagine what it could do to conventional glasses or unprotected eyes.

Gloves are another item that protect against meteorological conditions, and they also soften the contact with the handlebars and prevent road rash on the hands in case of a fall. Bicycling gloves usually are quite thin; they usually have padding in the palm, and the fingers may be covered by a finer weave or uncovered. Gloves must serve their protective function and simultaneously allow good mobility, and provide a good grip on the handlebars, the brakes, and the gear shifters. They should never be thick or capable of slipping off the handlebars, and wool gloves or other winter gloves with characteristics that are unsuited to riding should be avoided. In addition, in some cycling sports, the rider runs a greater risk of falling, as in cyclocross and mountain biking, while, in other types, such as road cycling, falls are much less common. This is one more factor to keep in mind in buying gloves with or without fingers, which may offer greater protection at the expense of comfort and vice versa.

Shoes need to be stiffer than the usual, which is the case with specialized cycling shoes, to avoid losing force as pressure is applied to the pedals, and they should not be equipped with laces—or else the laces should be protected—because they could get caught in the chain and cause the cyclist to crash. Obviously, cyclists can adjust their clothing and accessories to their level of involvement and apply these recommendations more precisely as their cycling increases in length and frequency.

Cycling shorts are an essential article of clothing, especially if you use a narrow saddle. Competition saddles tend to be narrow and to have little padding, because, above all, they are intended to be light and to promote efficient pedaling. This means that the contact points on the saddle experience lots of pressure and that there is constant friction. With their built-in padding, cycling shorts make cycling more pleasant and the effects more bearable.

Nutrition. Finally, and especially if you ride long distances, it is highly advisable to carry a full water bottle to avoid dehydration, especially in the summer and on hot days, some food rich in carbohydrates, and, of course, a phone with a charged battery. These recommendations can keep cycling from turning into an ordeal, and keep an unforeseen event from becoming anything more than a good story to share with friends.

All cyclists need a minimum of accessories in order to make their sports safe and enjoybale.

Biomechanics

In order to understand the demands that cycling places on your body, you first need to know the forces that are applied at different moments in pedaling and how they are produced, presuming that the bicycle has already been adjusted properly. You also need to keep in mind that it's not just the lower limbs that exert force, and that not all forces involve movement, because maintaining your position on the bicycle also constitutes important isometric work for various muscles of the upper body, neck, and upper limbs.

In addition, the parameters that are presented below cannot be applied in the same way to the various cycling sports, and that is also the case with adjustments. Even though the values are very similar in most cases, it should be understood that the characteristics of a cyclocross (BMX) bike and a racing bike are vastly different, to the same extent as the characteristics of the cyclists and their performance in the various sports.

Every person must determine which information is useful and which is not, because sports are not just a physical endeavor, but also mental and analytical. You always need to consider the reasons for things, think about the principles on which the advice you get is based, and take nothing at face value without prior reflection.

Now, we will provide some information applicable mainly to athletes who use racing and mountain bicycles, as well as to those who use bicycles in sports that have significant topographical variations.

DEMANDS ON THE LOWER BODY

The lower body is responsible for pedaling. This dynamic force produces a circular, cyclic movement by means of which you make the bicycle move forward. Of course, pedaling is what requires the greatest effort in cycling, so the lower body is very important, and you need to pay attention to it, not just while you are on the bicycle, but also on rest days and in the types of preparation you do that don't involve two wheels.

Pedaling is often divided into four phases that describe in simplified fashion the actions that you perform on the pedals throughout the cycle. This division is not sufficient for understanding the mechanics of pedaling, but it clarifies the basic actions that it involves. As with most divisions of an athletic movement into its various phases, it is appropriate to recall that these phases are fluid and imprecise, so they do not have a precise start and an exact end. Rather, the end of one phase overlaps the start of the next one; therefore, at certain times of the cycle, it is impossible to state precisely whether we are in one phase or another. The important thing is to have a general vision of the overall technical movement.

A BMX bike is very different from the standard bicycle in terms of its dimensions and structure.

A racing bike is used as the reference for most studies on the biomechanics of pedaling.

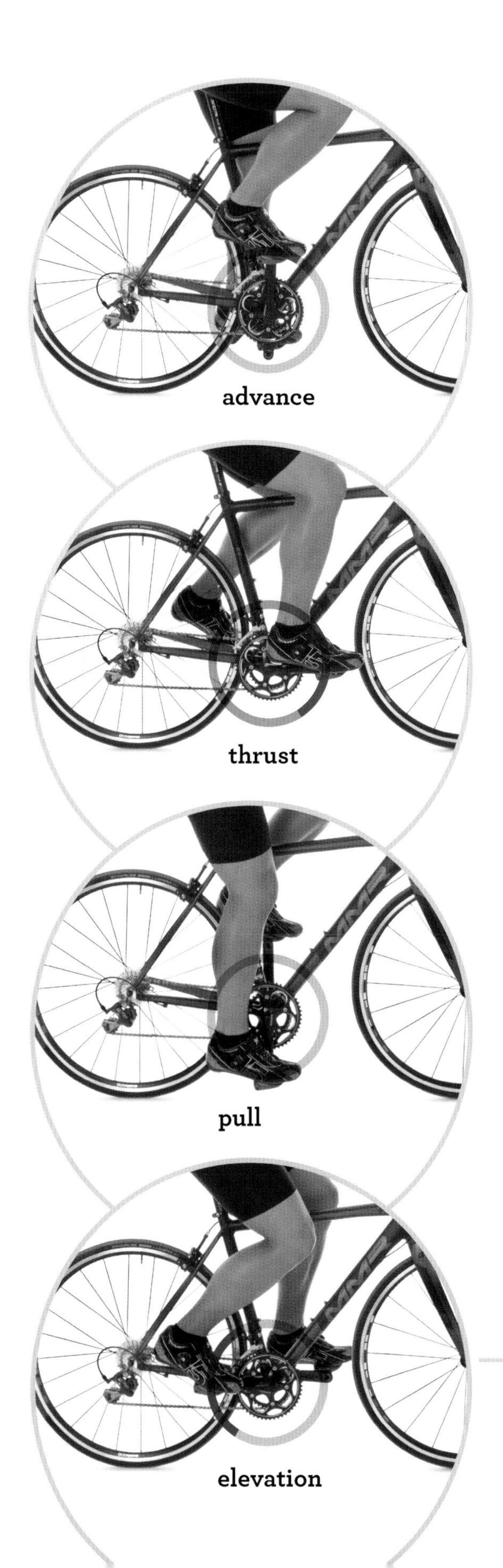

Advance: This is the part of the pedaling cycle where the foot is at its highest point. At this part of the cycle, the foot is moving forward, and, although this is the start of the moment of greatest force, it is accomplished gradually, coinciding with the end of the maximum force with the opposing limb, with the pedal located at the lowest point of the cycle.

Thrust: This is the part of the pedaling cycle that follows the advance and moves the pedal downward. This is the phase that requires the greatest force, and the one in which you move the pedal from high to low. This phase is the one where the greatest pressure is applied, and the one that is most responsible for moving the bicycle forward, so you need to pay particular attention to the muscles it involves.

Pull: This is the part of the cycle in which the pedal is at its lowest point and is moving rearward. It is the transition between the thrust and the elevation, or recovery, of the pedal, and the level of muscle force required declines as the effort exerted by the opposite leg increases during its advance phase.

Elevation: This is the part of the pedaling cycle in which the pedal rises; it requires very limited intensity, because the opposite leg is in the thrust phase and bears the brunt of the effort. Although there are as many ways to pedal as there are cyclists, and sometimes the elevation is done very actively and with lots of power, in general, this part of the cycle should be gentle in order to create efficient pedaling.

The advance, thrust, pull, and elevation phases constitute the pedaling cycle.

In order to analyze pedaling in greater detail now that we understand the basic actions required in its four main phases, we will analyze the muscle groups that come into play in each of these phases. The action of a given muscle group is not limited to one phase, but rather it can participate to a greater or lesser degree in several of them, and experience variations in intensity. The following chart will help us gain a clearer understanding of the pedaling phases in which the various muscle groups take part.

Hip Extensor Muscles: The most important muscle among the hip extensors is the gluteus maximus. Its significant role in pedaling, especially in the thrust and pull phases, means that gluteal strain is a common problem among cyclists. Other contributors to hip extension, but to a lesser degree, are the semitendinosus, semimembranosus, the long head of the biceps femoris, and, finally and to a much lesser extent, the gluteus medius and the adductor magnus.

Extensor Muscles of the Knee: These muscles, along with the hip extensors and the plantar flexors of the ankle, take on most of the work load and the force of pedaling. They play a crucial role in the advance and thrust phases, and they are the ones that make up the quadriceps femoris.

Plantar Flexor Muscles of the Ankle: The main muscles responsible for plantar flexion of the ankle, and the ones that perform a good deal of work, especially in the thrust and pull phases, are the gastrocnemius and the soleus, although to a lesser extent the tibialis posterior, the peroneus muscles, the flexor digitorum longus (pedis), and the flexor hallucis longus are also used.

Hip Flexor Muscles: These play a lesser role than the ones previously mentioned, because they are used mainly in the elevation and advance phases which involve less intensity. The muscles that are responsible for hip flexion are first, in terms of importance, the iliopsoas and the iliac, and then some others whose involvement is much less significant, such as the sartorius, the tensor fasciae latae, the pectineus, the rectus femoris, and the short and long adductors.

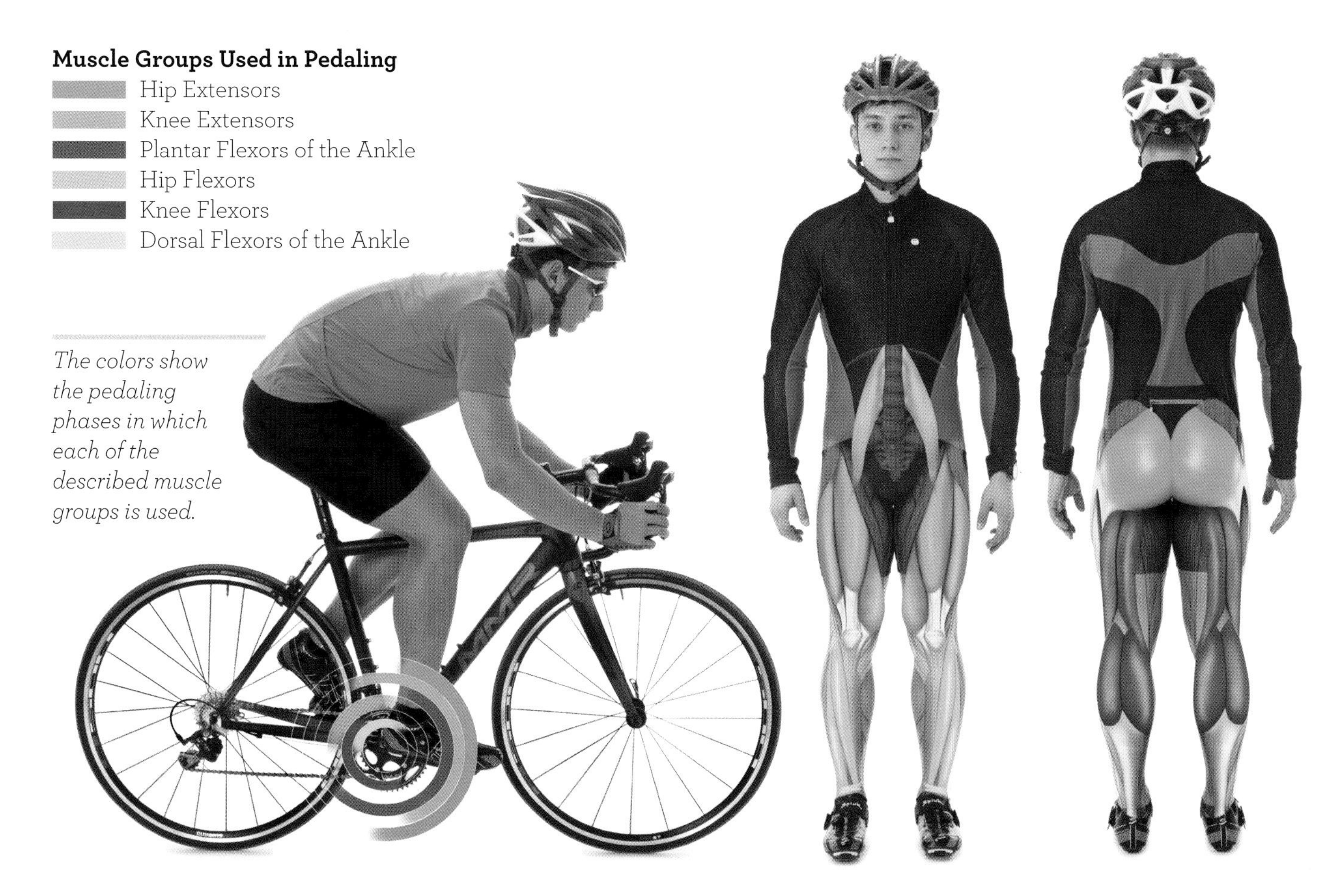

The colors show the pedaling phases in which each of the described muscle groups is used.

The requirements on the various muscle groups vary as a function of the type of pedaling used.

Knee Flexor Muscles: These muscles, like the hip flexors and the dorsal ankle flexors, likewise are not subjected to great tension, because their action is centered mainly on the pull and elevation phases. The muscles that play a greater role in knee flexion are the biceps femoris, the semitendinosus, and the semimembranosus. Other muscles also contribute, but to a lesser degree, such as the gracilis, the sartorius, the popliteus, and the gastrocnemius.

Dorsal Flexor Muscles of the Ankle: These are used mainly in the elevation and advance phases, and they are first of all the tibialis anterior, and then following in importance, the third peroneus, the extensor digitorum longus (pedis), and the extensor hallucis longus.

The action of these muscle groups can change its timing and intensity—and not only for reasons attributable to an individual cyclist's pedaling style. They may also experience variations due to pedaling positions. This involves not only certain adjustments, such as the saddle height, but also the specific timing of the event. Standing on the pedals involves variations in the muscle work as compared to pedaling while seated. This means, for example, that you are able to apply greater force to the pedals if you get out of the saddle to execute a breakaway, sprint, climb a steep hill, or simply change pedaling during a long climb. This variation in muscle work in changing from one position to another can provide greater strength through better use of muscle action and one's body weight; but, oftentimes, with great strength expenditure, there is a corresponding increase in fatigue. As a result, you must pedal intelligently. Another important element to be considered is whether or not your feet are locked onto the pedals. Clip-in pedals make it possible to perform some mildly active pull and elevation phases, whereas a conventional pedal without clips allows you only to push on it, but not to pull on it.

These comments provide some guidelines for improvements in your pedaling. By knowing at what point each of the muscle groups is working, you can deduce why one of them is prone to recurring or premature fatigue. For example, muscle strain in the tibialis anterior could be the result of an excessively energetic elevation phase in which you pull the pedal upward forcefully, with the resulting excessive workload for the muscles that are responsible for dorsal ankle flexion.

UPPER BODY DEMANDS

The upper body is often overlooked in cycling, because it does not contribute to driving the bicycle forward. But the fact that it is not the motor that generates the movement does not mean that it has nothing to do with cycling, that it is not important, or that you can leave it out of your considerations. A failure of any part of the upper body can be as crucial or limiting in an event as a muscle strain in the lower body.

You must not forget that the upper half of the body is in charge of holding up part of your body weight, controlling direction, providing you with a clear vision of the road and your rivals, applying the breaks and changing gears, holding an aerodynamic body position, and balancing the bicycle in a sprint.

Often, the muscles of the upper body that are used in cycling act isometrically and over time, so, at first glance, it may seem that they do not exert effort or get tired; however, even though their work is less obvious and most often performed in the background, they certainly are important.

Many cyclists have experienced pain or discomfort in the back and shoulders, which they could have avoided by varying their upper-body position or by going to the gym to train the relevant muscles.

The main upper-body muscles used in cycling.

Other muscle groups used in cycling

- Extensor Muscles of the Neck and Head
- Upper-body Extensors
- Shoulder Fixator Muscles
- Elbow Extensors
- Wrist and Finger Flexors

So let's take a look at the muscle groups that are surely of vital importance to cycling, because they contribute to maintaining your position on the bicycle, even though they are not used directly in pedaling.

Extensor Muscles of the Neck and Head: In cycling in general, and, to a greater or lesser extent, as a function of the specific cycling sport, the torso leans forward toward the handlebars. With a neutral position of the head and neck, this upper-body lean leaves the cyclist gazing at the ground, or in the best case, at the front wheel. Thus, in order to keep the cyclist's gaze directed ahead, onto the road, the track, or the selected line of travel, wherever it may lead, the cyclist must extend the neck and head. This neck position is held throughout most of the ride, and the extension diminishes when the cyclist stands on the pedals and increases when the cyclist adopts a particularly aerodynamic position on the bicycle. This musculature performs isometric work that lasts throughout the ride, so it needs to be developed and looked after. The most important muscles in holding and extending the neck and head are the splenius capitis and cervicis, the semispinalis capitis and cervicis, the trapezius, the rectus capitis posterior major and minor, the obliquus capitis superior and inferior, and the longissimus capitis and cervicis.

Extensor Muscles of the Upper Body: The work of these muscles is directly related to the position you assume on the bicycle. When it is more erect, on a leisure bicycle, for example, it takes less work from the muscles that are responsible for upper-body extension to keep your back straight. When you lean forward for a more aerodynamic position, the upper body tends to bend, so the muscles that counteract this tendency have to work harder. When the cyclist keeps a sharp bend in the upper body, as often happens in track racing, the musculature is subjected to tension, so in addition to maintaining an optimal position on the bicycle, it becomes necessary to strengthen this musculature and make it flexible so that it can withstand the various circumstances it may have to deal with in cycling.

It's a good idea to keep the back as flat as possible while on the bicycle.

Shoulder position may vary according to changes in pedaling positions.

The muscles that are responsible for extending the upper body and holding the position on the bicycle include the thoracic and lumbar iliocostals, the longissimus thoracis, the spinalis thoracis, the semispinalis thoracis, the musculi rotatori breves and longi, the musculi multifidi, and the cuadratus lumborum.

Shoulder Fixator Muscles: Along with the elbow, the shoulder is the joint that determines the degree of upper-body lean toward the handlebars, so it contributes to maintaining the cyclist's position on the bicycle, to the stability and the movement of the handlebars, and to the absorption of vibrations and impacts from irregularities on the ground transmitted through the handlebars from the front wheel. In large measure, the shoulder fixator muscles are also the ones that are responsible for balancing the bicyclist while standing on the pedals. The main muscles in use here are the deltoids, the pectoralis major, the latissimus dorsi, the teres major and minor, the supraspinatus, the infraspinatus, and the subscapularis.

Elbow Extensor Muscles: These muscles remain active during much of the cycling motion, because the elbow is held between total flexion and extension, which keeps the cyclist from collapsing onto the handlebars and makes it possible to deal with various situations in racing. This continuous effort may vary when pedaling on the flat with the elbows completely straight, or with the use of certain supports added to the handlebars, as in some track and triathlon events. Aside from these exceptions, sustaining the elbow position falls mainly to the triceps brachii and the anconeus, although the latter plays a largely token role.

Flexor Muscles of the Wrist and the Fingers: The grip on the handlebars must be firm and secure, so the fingers need to exert continuous tension, to which is added the momentary force required to use the brake levers and the gear shifters. In addition, resting the hands on the handlebars tends to cause complete extension of the wrist, produced by the effect of the upper-body weight. To maintain optimal health in this joint, you need to move it regularly by changing your grip and to avoid keeping it always in maximum extension, which can cause compression of the ulnar and medianus nerves, with the attendant negative effects that run the gamut from numbness in the hands and fingers to pain or momentary paralysis. For all these reasons, it is advisable to keep the muscles responsible for wrist and finger flexion in the best possible condition for strength and flexibility. Although there are many muscles that perform these functions, the most important ones are the flexor carpi radialis, the palmaris longus, the flexor digitorum superficialis, and the flexor digitorum profundus.

Regularly changing hand position on the handlebars helps alleviate pressure in the wrists.

EXTREME
RITCHEY
SCHWALBE

NECK AND UPPER BODY STRETCHES

NECK STRETCHES

Contrary to appearances, the neck is one of the areas that suffers and works the hardest in cycling, especially during long road races, whether single-day or stage races. The cyclist's inclined position on the bicycle, which carries the greatest aerodynamic efficiency, requires sustained extension of the head and neck in order to keep his or her gaze directed forward. This sustained effort entails lots of work, and often leads to straining of the muscles used to extend the head and the cervical vertebrae. In addition, turning the head rearward to locate at a glance a companion or a competitor involves additional work in lateral inclination and rotation of the head and neck, a function that in many cases the same extensor muscles perform when they contract unilaterally. Therefore, it is useful to know the main muscles that perform these tasks and to keep them in the best possible condition through strengthening and stretching exercises.

SCALENES

There are three different scalene muscles with fairly similar functions:

Anterior scalene: This originates at the anterior tubercles of the transverse apophyses of vertebrae C3 to C6, and it inserts at the tubercle of the anterior scalene located on the first rib.

Middle scalene: This originates at the posterior tubercles of the transverse apophyses of vertebrae C2 to C7, and it inserts at the upper surface of the first rib.

Posterior scalene: This originates at the posterior tubercles of the transverse apophyses of vertebrae C5 to C7, and it inserts at the outer face of the second rib.

When they act bilaterally, their function is to raise the first and second ribs or to bend the cervical vertebrae, and when they work unilaterally, they bend the latter to the side.

LEVATOR SCAPULAE

This muscle originates at the transverse apophyses of vertebrae C1 to C4, and it inserts at the upper angle of the scapula. It moves the scapula upward and in a medial direction, and it bends the cervical vertebrae to the side.

SPLENIUS CAPITIS

This originates at the nuchal ligament and the spiny apophysis of vertebrae T3 to C7. It inserts at the third lateral of the upper nuchal line of the occipital bone and at the mastoid apophysis of the temporal bone. Its bilateral contraction produces extension of the head and the cervical vertebrae, and its unilateral contraction bends and rotates the head and neck to one side.

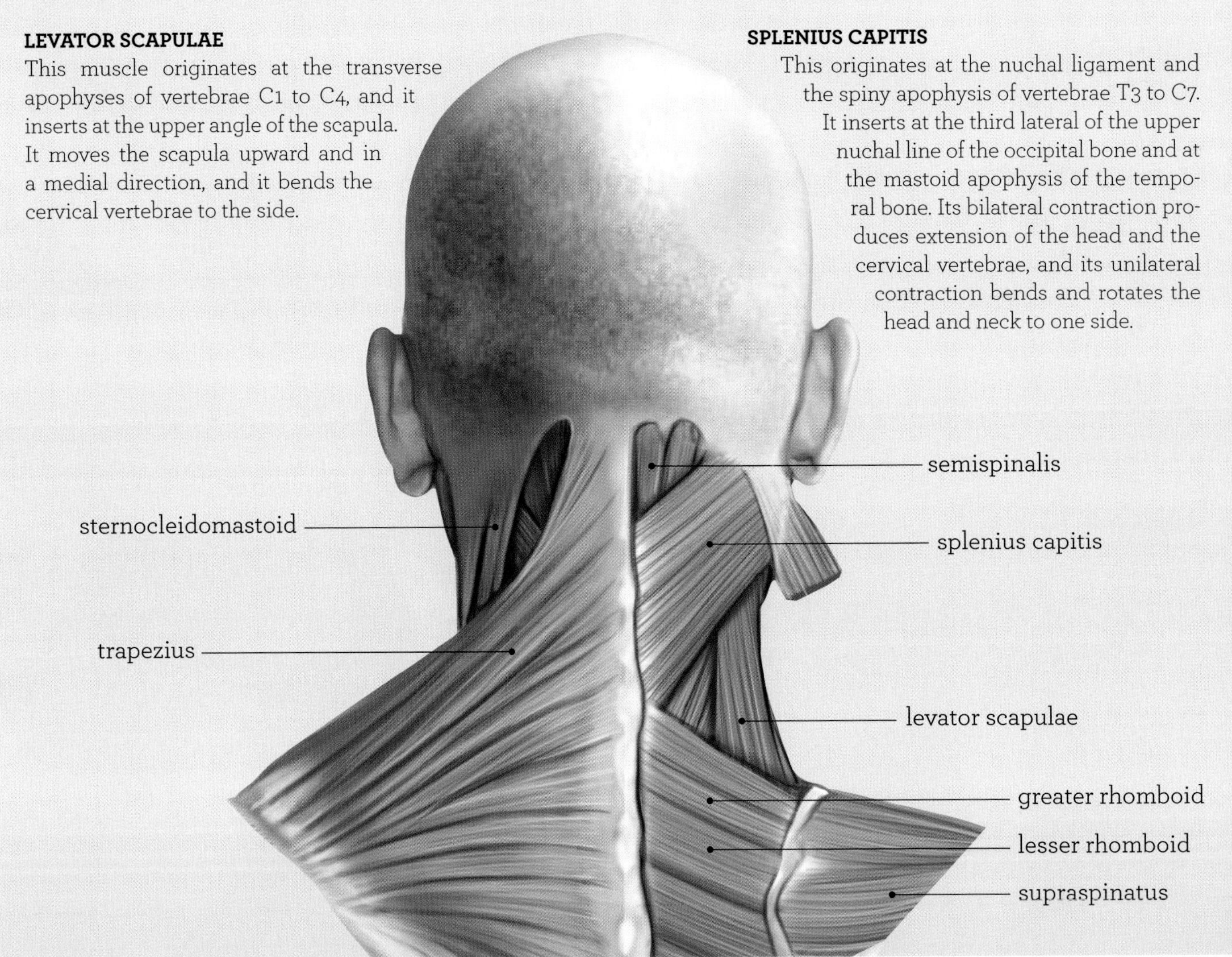

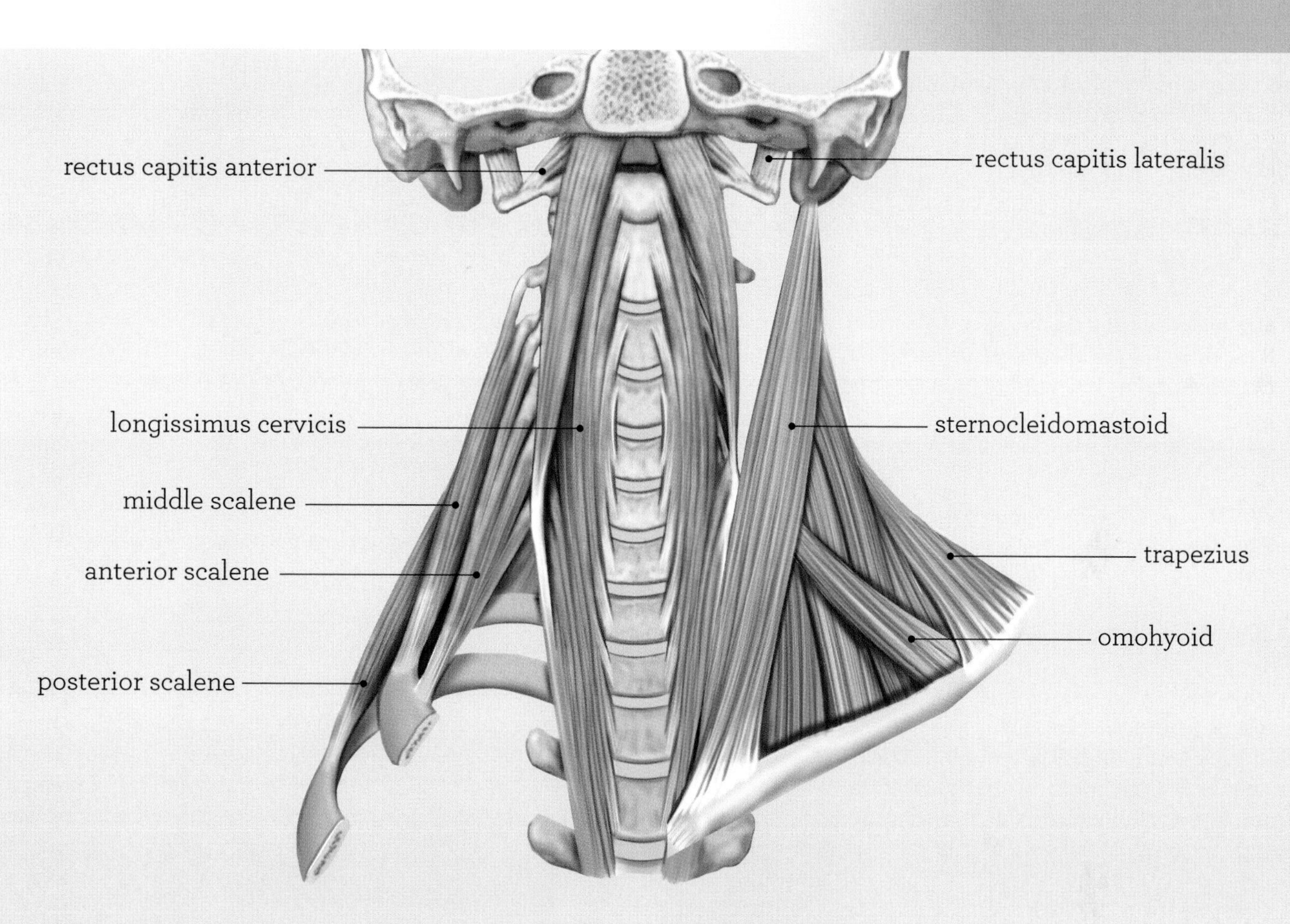

SPLENIUS CERVICIS

This originates at the spiny apophyses of vertebrae T3 to T6, and it inserts at the transverse apophyses of vertebrae C1 and C2. Its bilateral contraction produces extension in the cervical vertebrae, and its unilateral contraction causes flexion and rotation of these vertebrae.

STERNOCLEIDOMASTOID

Its sternal end originates at the manubrium, and the clavicular end at the middle third of the clavicle. It inserts at the mastoid apophysis of the temporal bone and at the upper nuchal line of the occipital bone. Its bilateral contraction causes extension of the head, and unilateral contraction rotates the head toward the opposite side while raising the chin.

SEMISPINALIS

The complete muscle covers the occipital bone to the transverse apophyses of vertebra T12. It is divided into three parts, the first two of which are of special importance in this section:

Semispinalis capitis: This originates at the transverse apophyses of vertebrae C4 to T7, and it inserts between the upper and lower nuchal lines of the occipital bone.

Semispinalis cervicis: This originates at the transverse apophyses of vertebrae T1 to T6, and it inserts at the spiny apophyses of vertebrae C2 to C5.

Semispinalis thoracis: This originates at the transverse apophyses of vertebrae T6 to T12, and it inserts at the spiny apophyses of vertebrae C6 to T4.

Its bilateral contraction produces extension of the head and the cervical and thoracic vertebrae, and its unilateral contraction inclines and rotates the head and the cervical and thoracic vertebrae.

TRAPEZIUS

Its origin is broad, from the upper nuchal line of the occipital bone, passing through the nuchal ligament and the spiny apophyses of vertebrae C7 to T12. It inserts at the lateral third of the clavicle, the acromion, and the spine of the scapula. In this section, the descending part of this muscle is of particular interest, because one of its functions is the extension of the head and the cervical vertebrae.

Neck Bend

START
Stand with your back straight, your feet slightly separated, and your arms at your sides. Look straight ahead and relax before you begin the stretch.

TECHNIQUE
Let your head drop forward by bending your cervical vertebrae. From this position, try to accentuate the neck and head bend, as if you were trying to touch your chin to your breastbone, while keeping your eyes on your toes. Hold this position for a few seconds, and you will quickly notice tension in the rear of your neck and the top of your thoracic vertebrae.

LEVEL	REPS	DURATION
BEGINNER	2	15 sec
INTERMEDIATE	3	20 sec
ADVANCED	3	25 sec

CAUTION

You should not pull your chin inward, but rather try to reach your breastbone with it by bending your head and neck. If you perform this movement as far as you can reach, regardless of how far that is, you will be doing the stretch.

INDICATION

For athletes who experience muscle tension in the back of the neck, especially road and track riders who spend long periods in training or competing on the bicycle, because of their position on it.

Self-assisted Neck Bend

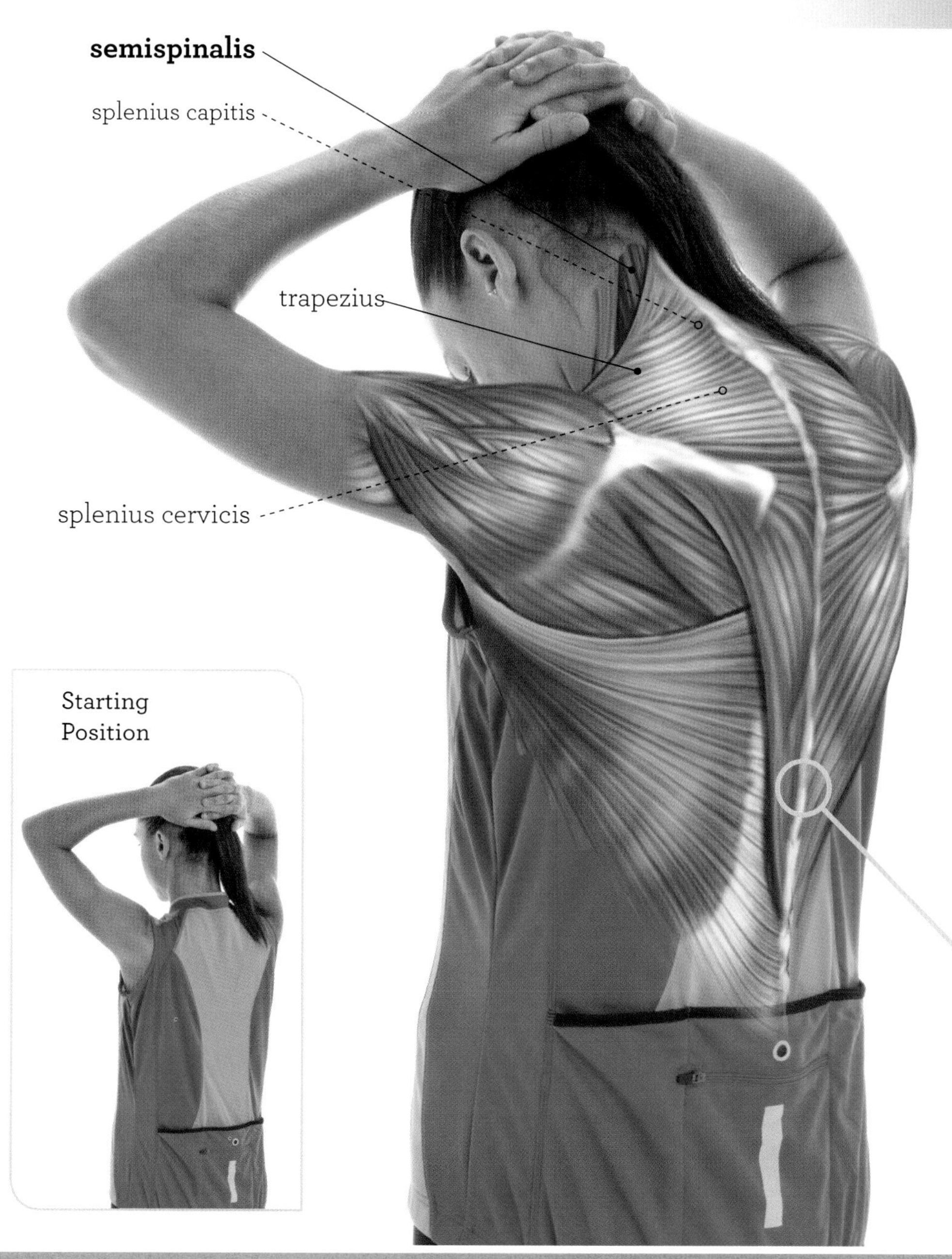

START
Stand with your back straight and look straight ahead. Place your hands behind your head and in contact with it, with your fingers locked together firmly.

TECHNIQUE
Pull your head forward and downward with both hands to produce a bend in your neck and head. When you look at your toes, you will feel the tension in the back of your neck due to the stretch in the semispinalis and other muscles.

Keep your back straight and look down at the floor.

LEVEL	REPS	DURATION
BEGINNER	2	15 sec
INTERMEDIATE	2	20 sec
ADVANCED	2	25 sec

CAUTION

Be careful when you apply pressure to your head, because the cervical vertebrae are very sensitive to force and traction applied abruptly or too intensely. If you experience any pain or discomfort, reduce the intensity of the pull or discontinue the exercise.

INDICATION

For athletes who experience muscle tension in the back of the neck, especially for cyclists who maintain an aerodynamic lean on the bicycle for extended periods, whether in competition or training.

Neck Bend and Rotation

START
Stand with your legs slightly apart, your back and neck erect, and look straight ahead. Your hands can be at your sides, or you can hold them behind your back.

TECHNIQUE
Lower and rotate your head toward the side opposite the stretch while keeping your gaze directed at the space near the outside of your foot. Try to extend the movement, as if you wished to reach the front of your shoulder with your chin. The tension produced by the stretch will become evident in the back of your neck and in the side opposite to which you are looking.

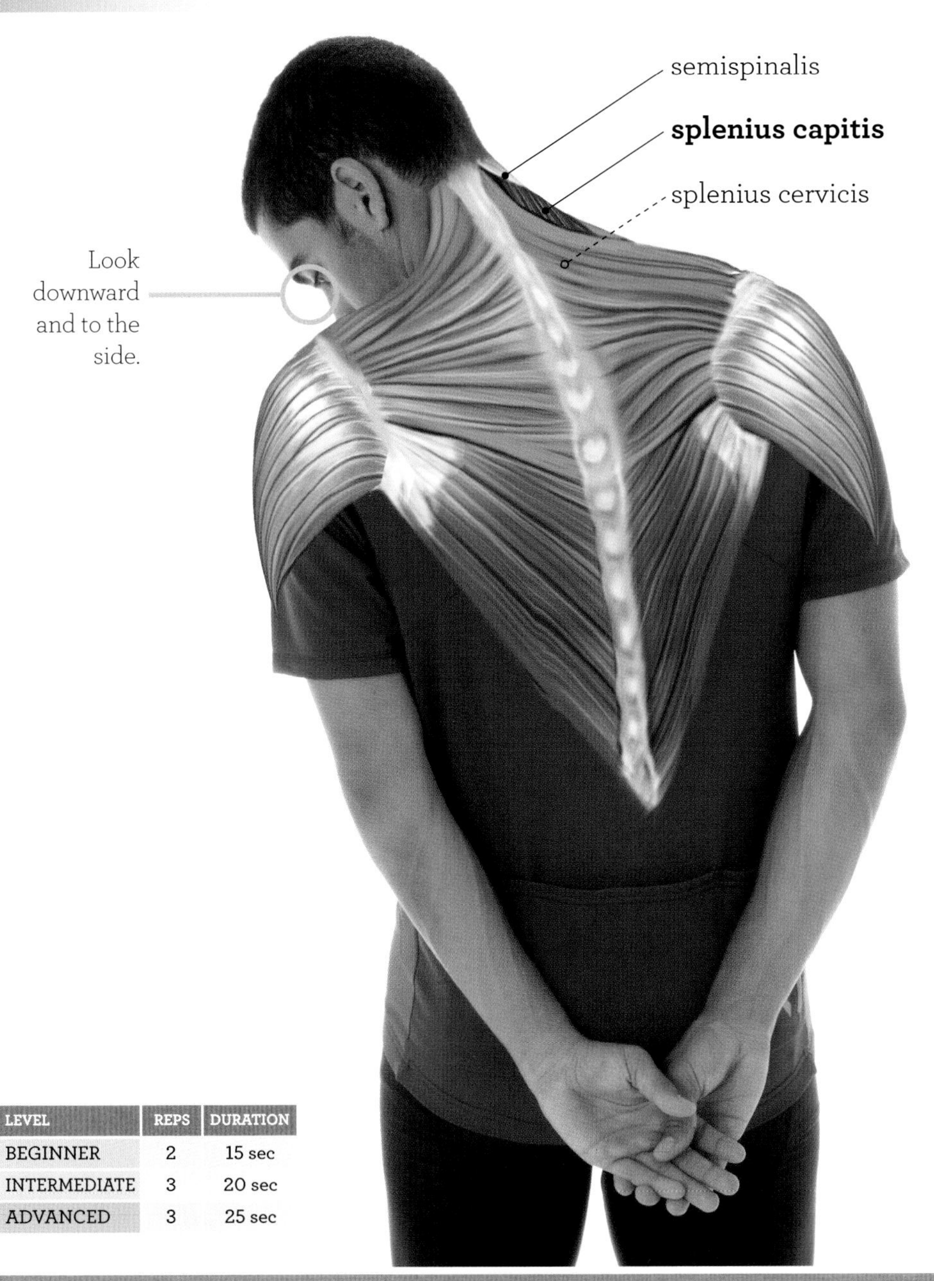

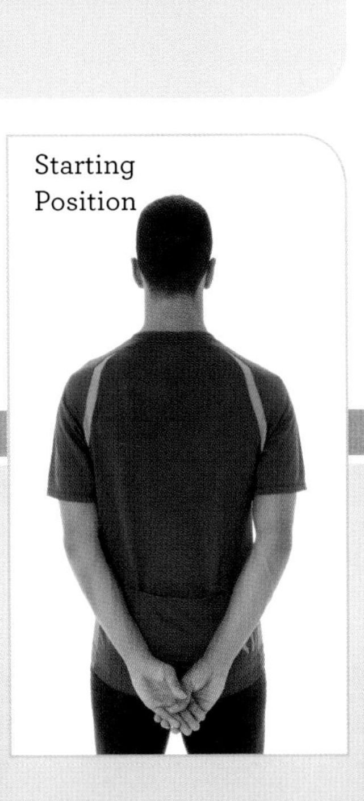

LEVEL	REPS	DURATION
BEGINNER	2	15 sec
INTERMEDIATE	3	20 sec
ADVANCED	3	25 sec

CAUTION

Avoid leaning your upper body or moving your shoulder toward your chin. These movements may give the false sensation of producing a greater range and stretch, but they contribute no effectiveness to the exercise.

INDICATION

For athletes who suffer from muscle tension in the back of the neck, and particularly for track and road racers, because of the increased activity of the neck and head extensor muscles in these events.

Self-assisted Head and Neck Rotation

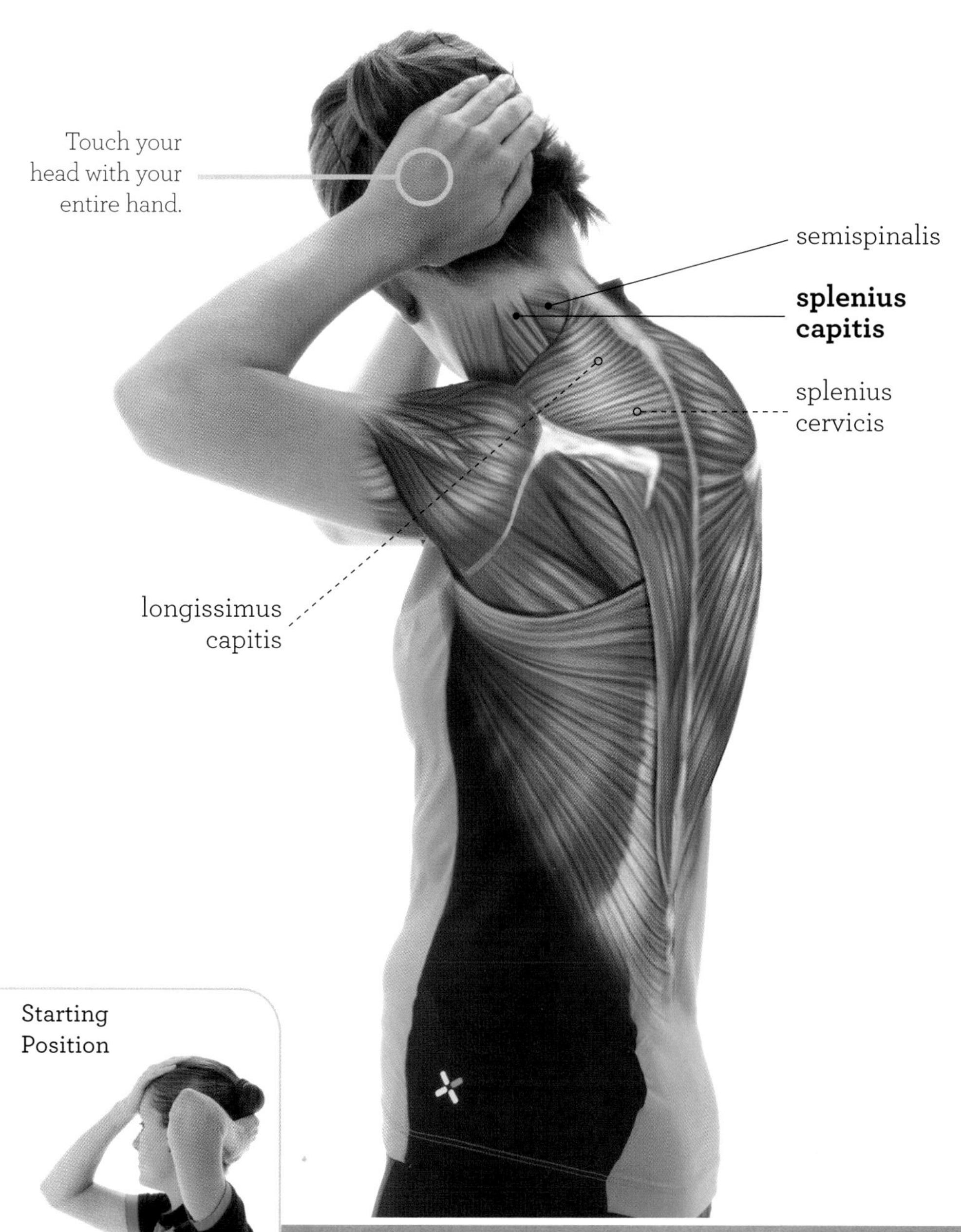

Starting Position

START

Starting from a standing or seated position, hold one hand on your forehead so that your palm touches your forehead and your fingers are on your head. Place the other hand on the back of your head at the level of your occipital bone.

TECHNIQUE

Bend your neck slightly as you use both hands to rotate your head toward the side opposite the stretch. Keep your gaze directed toward your side and slightly downward at the moment of maximum stretch, and you will feel the tension in the back of your neck, near the insertion of the splenius muscle.

LEVEL	REPS	DURATION
BEGINNER	2	15 sec
INTERMEDIATE	2	20 sec
ADVANCED	3	20 sec

CAUTION

Be gentle and gradual in starting the rotation and don't use much force; the area you are manipulating during this exercise is very sensitive.

INDICATION

For cyclists whose position on the bicycle involves pronounced forward lean, because of the need to maintain an aerodynamic position like the one used in track cycling and time-trial events. Also for riders who compete on roads, in long races, as well as for any athletes who experience muscle tension or strain in the back of the neck.

Self-assisted Lateral Neck Bend

START
Start in a standing position with your back and head erect. Raise one arm and place your hand on the opposite side of your head, with your forearm above it. Your fingers will point toward the ground, with your fingertips touching your ear.

TECHNIQUE
Gently and gradually pull on your head to cause it to lean so that your ear approaches your shoulder. When it gets close, you will feel the tension in the opposite side, from the head to the shoulder. You can place the inactive arm on your back to increase the intensity of the stretch.

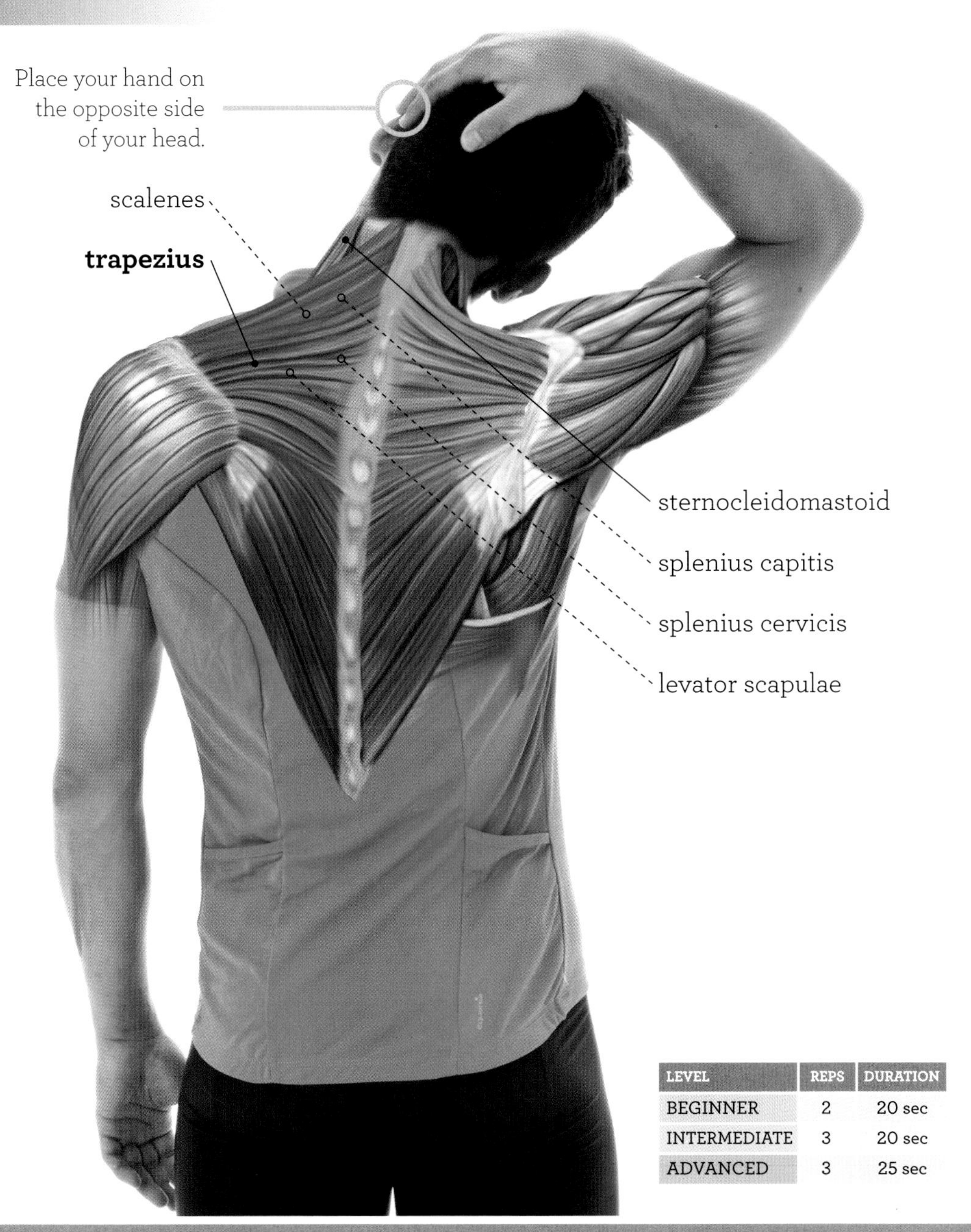

LEVEL	REPS	DURATION
BEGINNER	2	20 sec
INTERMEDIATE	3	20 sec
ADVANCED	3	25 sec

CAUTION
Perform the pull gently and gradually, avoiding sudden pulls and excessive force, given the great sensitivity of the area being manipulated.

INDICATION
For cyclists who take part in road events, whether in stages or on a single day, especially long ones. Also for cyclists in other disciplines and athletes in general who experience muscle tension or strain in the back of the neck.

Rear Neck Pull

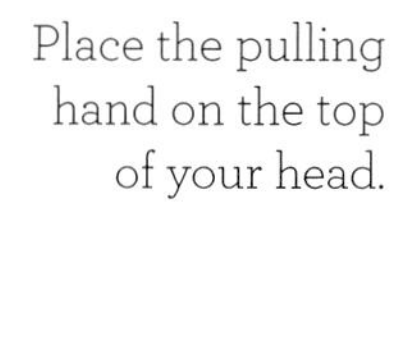

Starting Position

START

Turn your head slightly toward the side opposite the stretch. Raise the arm on the side toward which you are looking and place your hand onto the top of your head, so that your fingers point downward and your forearm is on top of your head. Even though your head is slightly turned, your neck and back should remain at a right angle to your body.

TECHNIQUE

Gradually pull on your head, causing your neck to bend and rotate so that your face comes close to your armpit. Do not pull with excessive force, and pay attention to your sensations, which should go no further than muscle tension in the back of your neck.

LEVEL	REPS	DURATION
BEGINNER	2	15 sec
INTERMEDIATE	2	20 sec
ADVANCED	2	25 sec

CAUTION

Keep your lumbar vertebrae straight and stop the exercise if you feel any discomfort or pain in these vertebrae. Remember that the neck is a very delicate area.

INDICATION

For cyclists, in particular ones who do track or road events, because of the strain produced in the extensor muscles of the head and neck.

UPPER BODY STRETCHES

In cycling, the front and back of the upper body perform very unbalanced work. The abdominal muscles are relaxed and relatively tight, while the back muscles are subjected to significant tension. If you envision a mountain biker, a track cyclist, or road racer, you will see that all of them bend significantly over the bicycle, but the position is much more exaggerated in the road or track racer. This involves sustained stretching and tension in the spinal erector muscles, which need to be strong enough to bear this work load and sufficiently flexible to avoid excessive discomfort in situations where the cyclist must adopt a more aerodynamic position, such as in a downhill race or a time trial. These factors, along with the degree of flexibility in other muscle groups, the distension of the vertebral ligaments, and the position on the bicycle, are largely responsible for the presence or the absence of pain, especially in the lower lumbar area of the back.

RHOMBOIDS

Even though there are both greater and lesser rhomboid muscles, they are arranged contiguously; in fact, they act as a single functional unit, and they are stretched in the same way.

Greater rhomboid: This muscle originates at the spiny apophyses of vertebrae T1 to T4, and it inserts at the middle edge of the scapula.

Lesser rhomboid: This muscle originates at the spiny apophyses of vertebrae C6 and C7, and it inserts at the middle edge of the scapula, above the insertion of the greater rhomboid.

The function of both muscles is to retract and raise the scapula.

LONGISSIMUS THORACIS

This muscle is one of the three parts of the longissimus muscle, which is completed by the longissimus capitis and cervicis, and it is one of the muscles that are referred to as the paravertebrals. It originates at the sacral bone, the middle third of the iliac crest, and the transverse apophyses of the lumbar vertebrae. It inserts at the second through the twelfth ribs and the transverse apophyses of the thoracic vertebrae. Their bilateral action extends the spine, and unilateral contraction bends it to one side.

ILIOCOSTAL

Even though there are three defined parts of the iliocostal muscle, namely the cervical, the thoracic, and the lumbar iliocostals, here we will focus on the latter two, because they are the ones that are located on the back of the upper body.

Thoracic iliocostal: This muscle originates at ribs seven through twelve, and it inserts at ribs one through six.

Lumbar iliocostal: This muscle originates at the sacral bone, the iliac crest, and the thoracolumbar fascia, and it inserts at ribs seven through twelve.

They are part of the paravertebral muscles, so their main function is extending the spine, although when they act unilaterally, they bend it to the side.

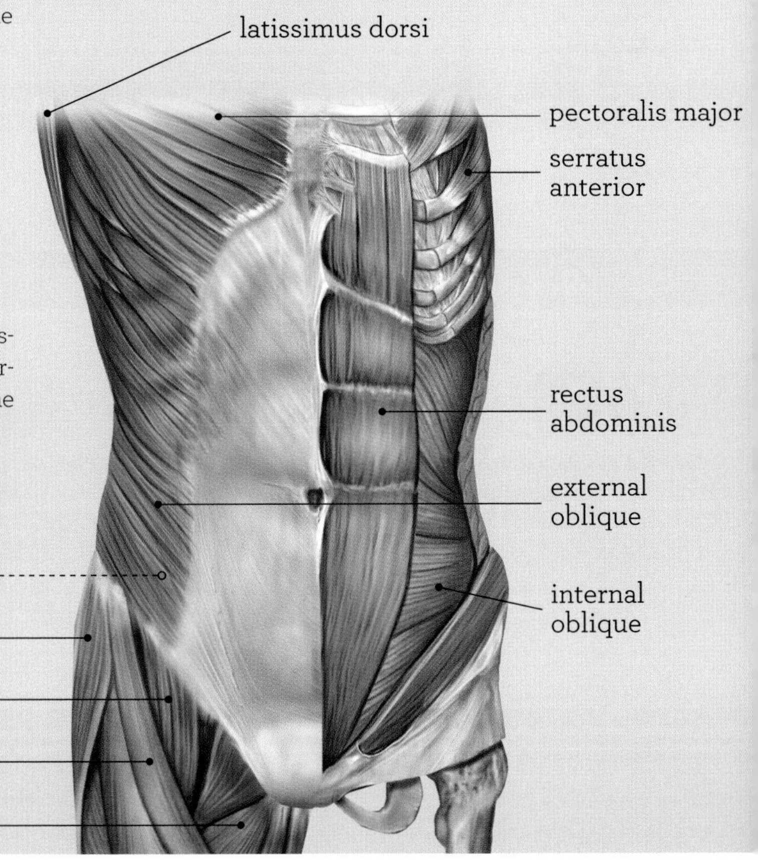

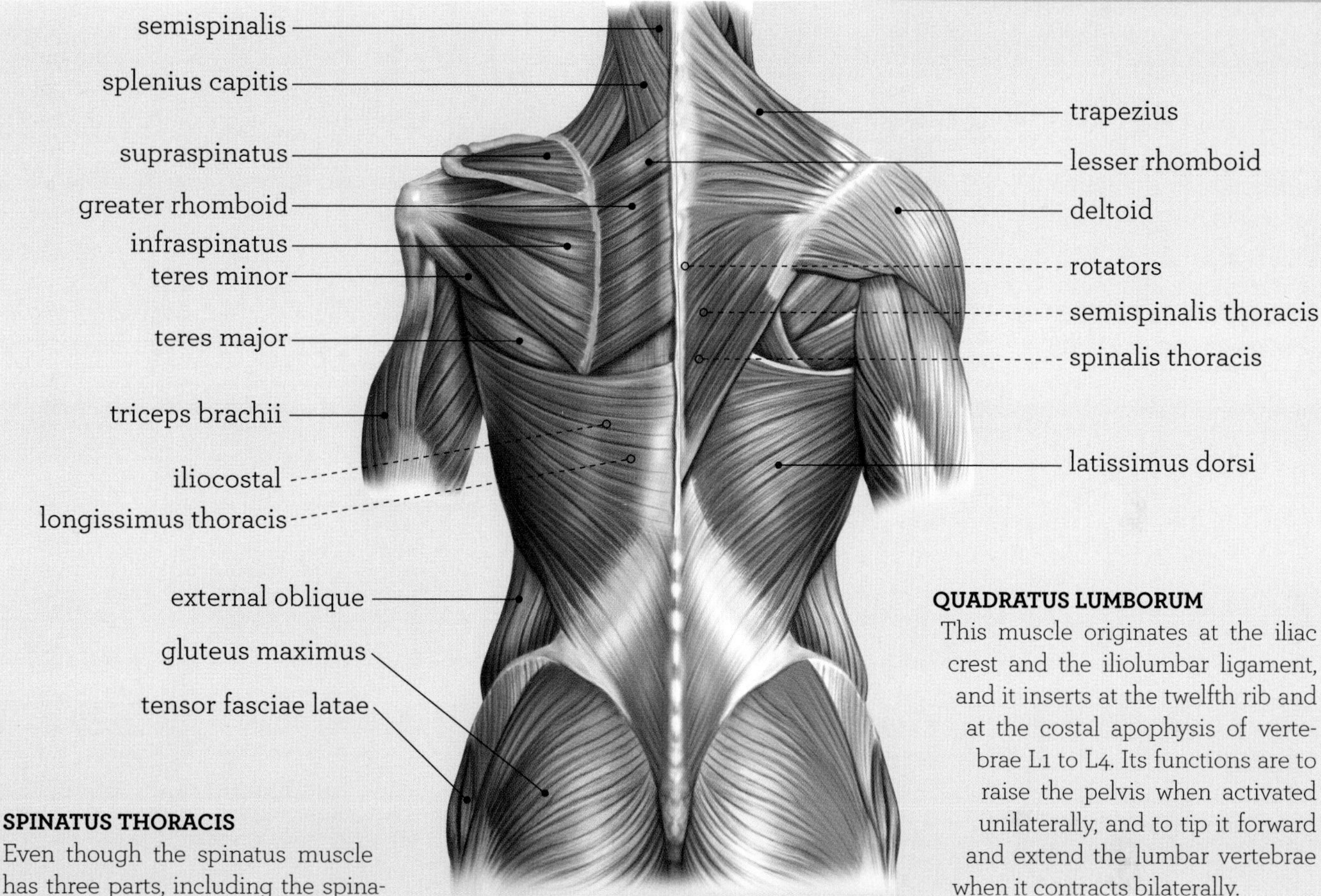

QUADRATUS LUMBORUM

This muscle originates at the iliac crest and the iliolumbar ligament, and it inserts at the twelfth rib and at the costal apophysis of vertebrae L1 to L4. Its functions are to raise the pelvis when activated unilaterally, and to tip it forward and extend the lumbar vertebrae when it contracts bilaterally.

SPINATUS THORACIS

Even though the spinatus muscle has three parts, including the spinatus cervicis and capitis, the spinatus thoracis is the one that acts exclusively on the back of the torso, so we will deal with it in this section. It originates at the spiny apophyses of vertebrae T10 to L3, and it inserts at the spiny apophyses of vertebrae T2 to T8. Their bilateral action extends the spine, and their unilateral action bends it to one side.

SEMISPINALIS THORACIS

The semispinalis muscle also has a part of the head and the neck, but as with the previous cases, we will focus on its thoracic part. The semispinalis thoracis muscle originates at the transverse apophyses of vertebrae T6 to T12, and it inserts at the spiny apophyses of vertebrae C6 to T4. As with the other paravertebral muscles, its bilateral contraction extends the spine, and its unilateral action bends it toward the side.

ROTATORS

There are short and long rotators, and their attachment points are in the spiny and transverse apophyses of vertebrae T1 to T12. They serve to extend and rotate the spine.

RECTUS ABDOMINIS

This muscle originates at the crest and the symphysis of the pubis, and it inserts at the xiphoid apophysis and at the cartilage of ribs 5 through 7. Its functions are bending the upper body and compression of the abdomen.

EXTERNAL OBLIQUE

This muscle originates at ribs five through twelve, and it inserts at the linea alba, the pubis, and the iliac crest. Its functions are the bending and the rotation of the upper body when used unilaterally, and the bending of the upper body and compression of the abdomen when it is used bilaterally.

INTERNAL OBLIQUE

This muscle originates at the thoracolumbar fascia, the iliac crest, and the inguinal ligament, and it inserts at ribs ten through twelve and the linea alba. When it contracts bilaterally, its action involves bending the upper body and compressing the abdomen, and when the contraction is unilateral, it bends and rotates the torso.

Seated Crossover Pull

START
Sit on a stool or a chair with no arms. Cross one leg over the other so that the outside of the ankle of the crossed leg rests on the thigh of the support leg. Using the hand on the side you will stretch in pronation, hold the raised foot so that the tips of your fingers grasp the outer edge of your foot firmly.

TECHNIQUE
Try to incline or move the center of your back rearward without letting go of or moving your grip hand, to produce abduction of the scapula. As you increase the pull, you will feel greater intensity of the stretch in the upper area of your back and the side of the pulling arm.

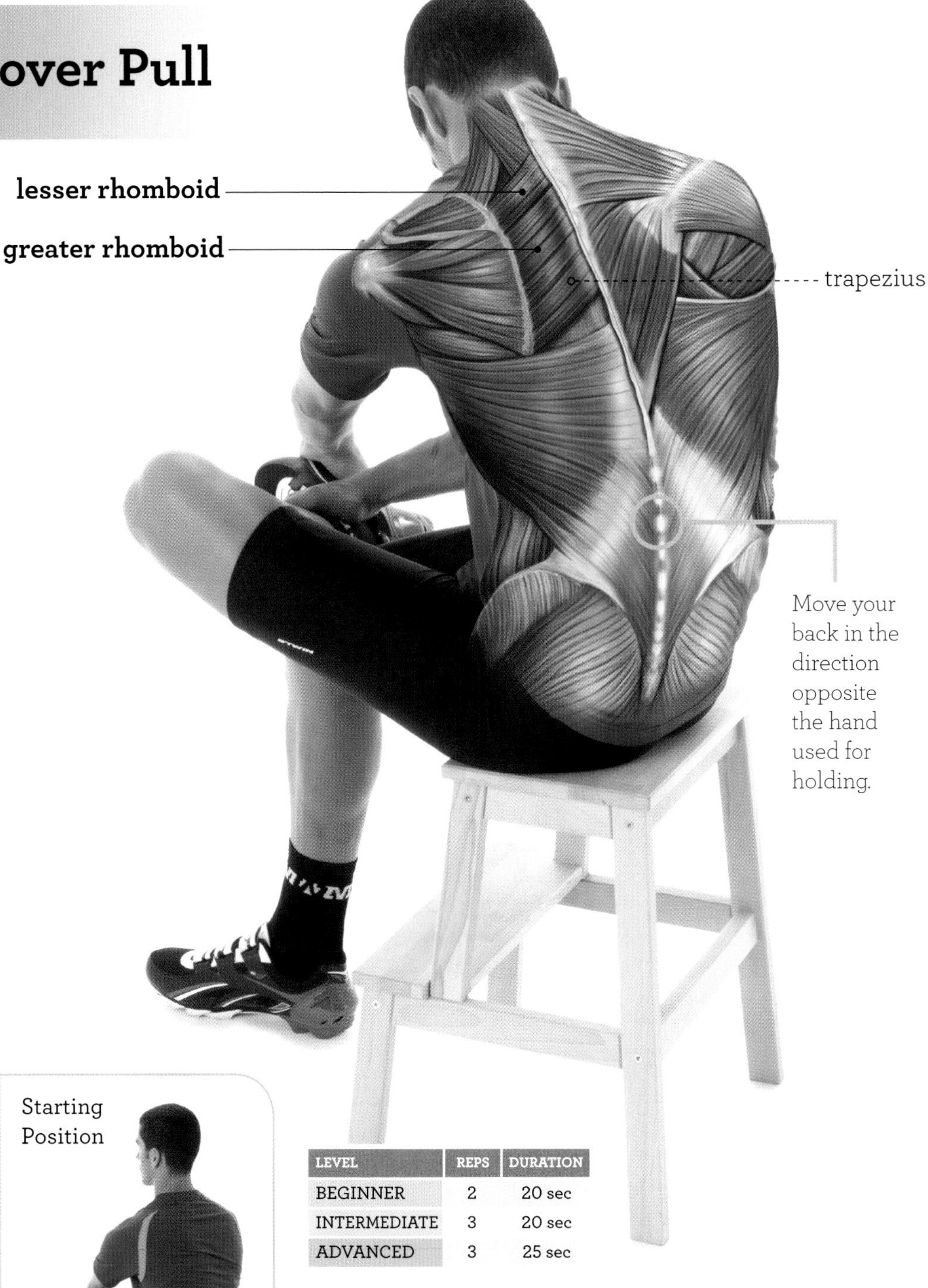

Starting Position

LEVEL	REPS	DURATION
BEGINNER	2	20 sec
INTERMEDIATE	3	20 sec
ADVANCED	3	25 sec

CAUTION

Relax the top of your back to allow abduction of the scapula and, consequently, the stretch.

INDICATION

For all cyclists, especially in sports such as cross-country cycling, downhill racing, and various bicycle motocross, or BMX, events, but also including bicycle touring, because of the more erect position on the bicycle, in which the rhomboids remain relaxed in a shortened position for most of the time.

Hug

deltoid

lesser rhomboid

trapezius

greater rhomboid

Make the center of your back the most prominent part while doing this stretch.

START

You can do this stretch either seated or standing. Your back and neck must be erect, as you look straight ahead. Cross your arms in front of your chest and place each hand on the opposite shoulder. In each successive set, alternate the arm that crosses over the other one in order to avoid asymmetry or imbalances in the stretch.

TECHNIQUE

Start from the position described, and without uncrossing your arms, try to reach the center of your back, the part between your shoulder blades, with each hand, so that your chest is compressed and the upper part of your back is slightly curved toward the rear. This will cause abduction of the shoulder blades.

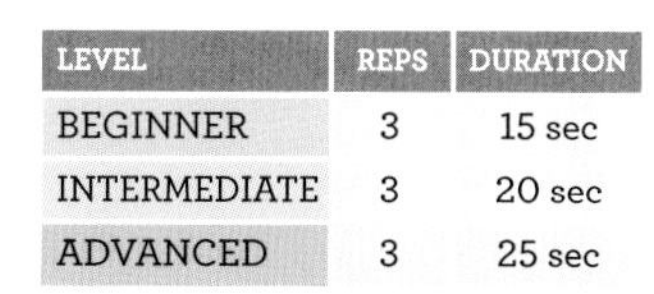

LEVEL	REPS	DURATION
BEGINNER	3	15 sec
INTERMEDIATE	3	20 sec
ADVANCED	3	25 sec

Starting Position

CAUTION

This exercise involves no substantial risk or complexity, although you probably will have to try hard to maintain the position of maximum stretch, because there is no anchor point involved.

INDICATION

For users of mountain bikes, recreational bicycles, and BMX bikes, which require a more erect position, in which the shoulder blades are in adduction and the rhomboid muscles are in a relaxed, shortened position for most of the riding or recreational time.

Squat and Leg Pull

START
Squat down with your feet slightly more than a hand's breadth apart. Slightly straighten your knees until your thighs are parallel to the ground and at an angle of about 80° with your legs. Place your arms beneath your thighs, as if you were hugging them, and use each hand to grasp the opposite elbow. Your chest will be in contact with or very close to your thighs.

TECHNIQUE
Slightly straighten your knees without letting go of your elbows so that the pull in your elbows increases along with the abduction of your shoulder blades, and the upper middle part of your back sticks out the most. This will cause tension in the rhomboid muscles on both sides of your back.

LEVEL	REPS	DURATION
BEGINNER	1	20 sec
INTERMEDIATE	2	15 sec
ADVANCED	2	20 sec

Starting Position

CAUTION

This exercise involves subjecting the lumbar area to a certain amount of tension, so try to keep your lumbar vertebrae straight to the extent possible, and skip this exercise if you feel or have previously felt any discomfort in the lumbar region.

INDICATION

For cyclists whose sports require a more erect position on the bicycle, and ones in which the rhomboid muscles are relaxed and shortened, as in downhill racing, cross-country cycling, or BMX. Also for bicycle tourists and regular users of recreational bicycles.

Seated Leg Pull

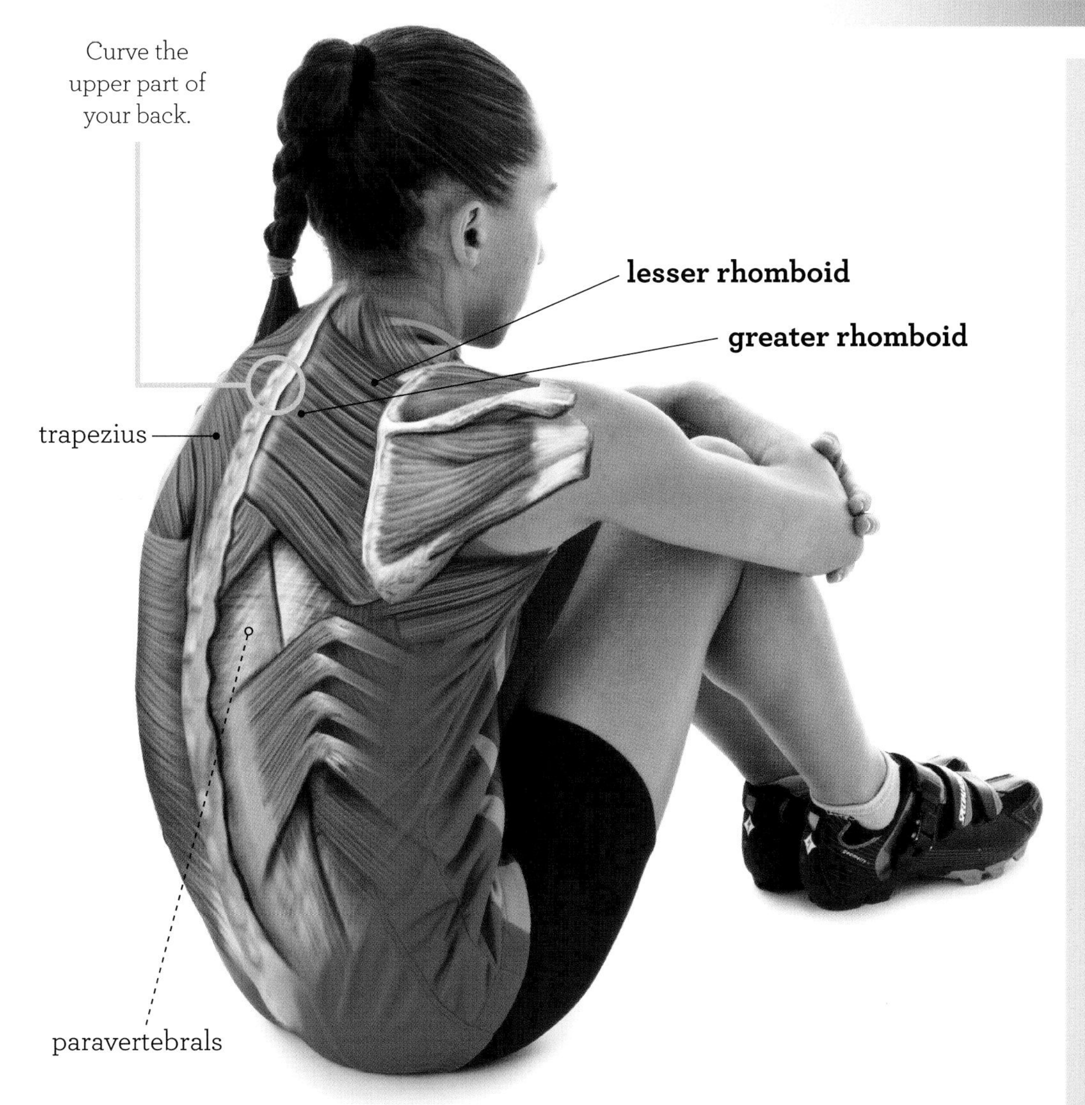

START
Sit down and bend your knees so that you are resting on your gluteus muscles and the soles of your feet. Cross your forearms in front of your knees and hold each elbow with the opposite hand. At this point, your back should be erect and your upper body may lean slightly forward.

TECHNIQUE
Try to move your thoracic vertebrae rearward without letting go of your elbows, so that your back bends and produces abduction of your shoulder blades. The tension will become apparent in the upper middle of your back as the stretch progresses. Remember that neither your support points nor your grip should move from the initial position.

LEVEL	REPS	DURATION
BEGINNER	2	15 sec
INTERMEDIATE	2	20 sec
ADVANCED	3	20 sec

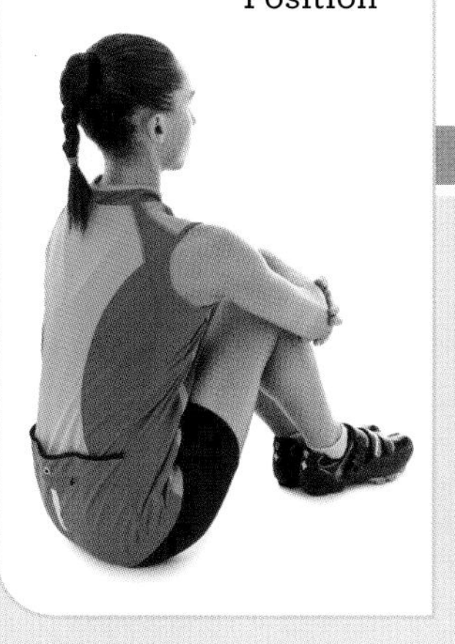
Starting Position

CAUTION
Even though the tension in the lumbar area during this exercise is much less than in the previous exercise, it is not entirely without risk, so pay particular attention to any irritation or discomfort in this area, and stop the exercise if you feel anything.

INDICATION
For riders of mountain, downhill, BMX, recreational, and hybrid bikes, and any other bikes that require a fairly erect posture in which the shoulder blades are in adduction.

Crossover Grip with Partner

START
Stand a short distance in front of a partner. Hold hands in a crossed position, so that your right hand holds your partner's left, and do the same with the left one. Your wrists will be crossed one over the other, with your elbows bent at an angle a little over 90°.

TECHNIQUE
While standing solidly on your feet, extend your arms toward the front so that you lean toward the rear, as your partner holds you and likewise leans rearward to provide a counterweight. Try to move the upper middle part of your back rearward to produce abduction in both shoulder blades.

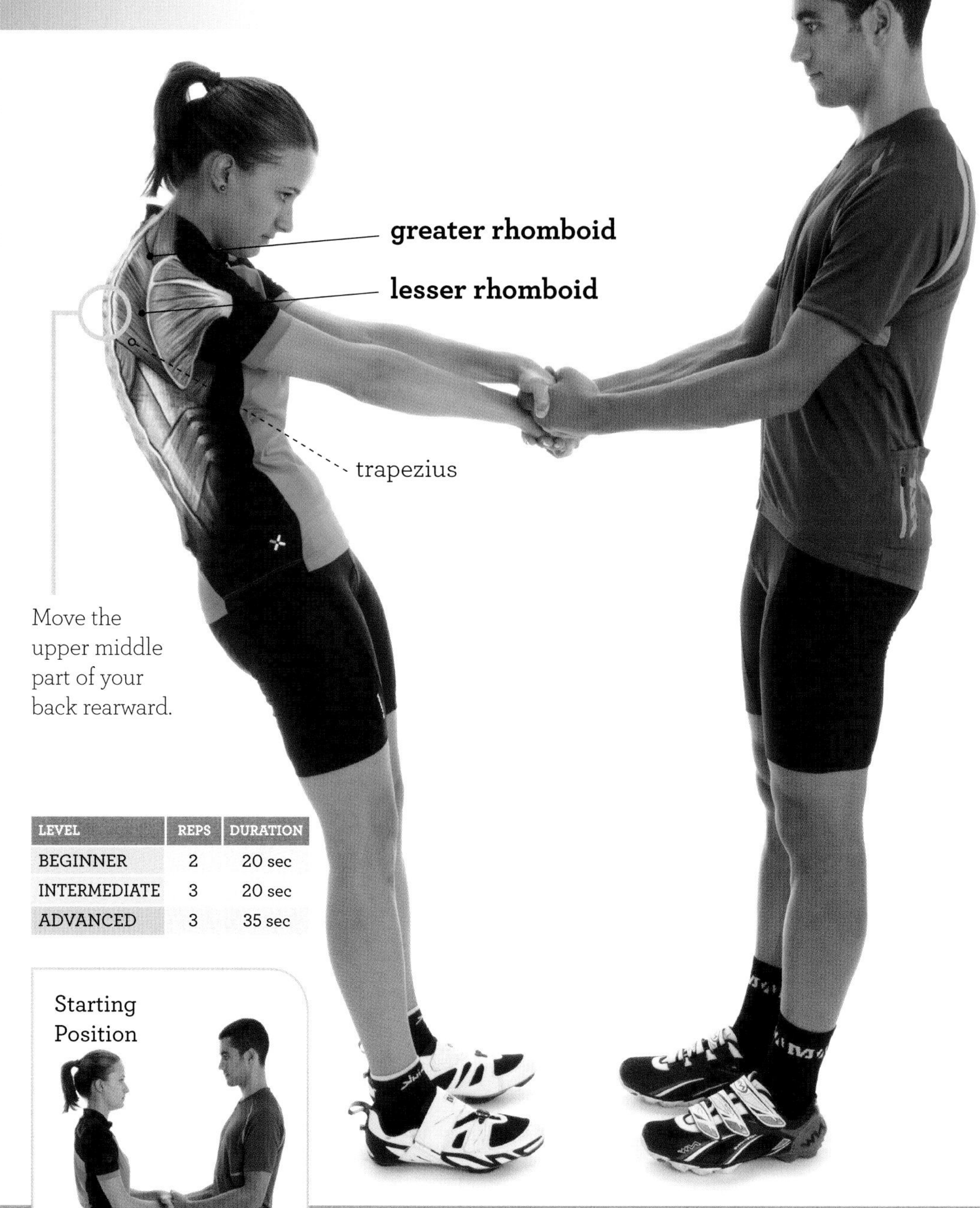

Move the upper middle part of your back rearward.

LEVEL	REPS	DURATION
BEGINNER	2	20 sec
INTERMEDIATE	3	20 sec
ADVANCED	3	35 sec

Starting Position

CAUTION

Lean to the rear slowly so that your partner can modify his or her position and act as an effective counterweight to maintain a good, balanced position.

INDICATION

For cyclists who ride mountain bikes frequently or intensively, recreational bikes, hybrids, and others that require a more erect position than racing bicycles do.

Cat Position

Starting Position

Keep your spine curved.

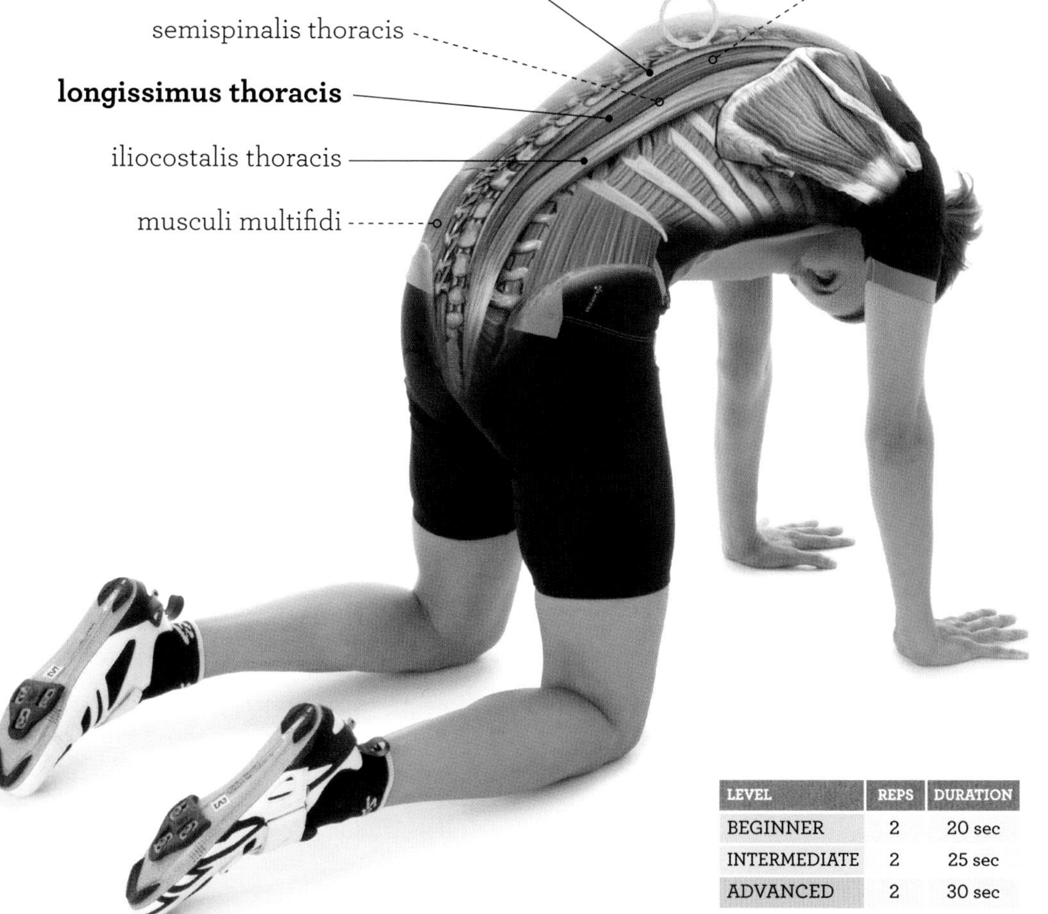

START

Get down on all fours, with your thighs and arms perpendicular to the floor. Your back must be straight; otherwise, you could upset the anatomical curves natural to the spinal column. The distance between your knees and your hands should be about shoulder width.

TECHNIQUE

Lower your head and arch your back like a cat. Try to achieve the greatest possible curvature and produce an adequate stretch in the spinal erector muscles. You may not feel the muscle tension from the stretch, but this does not mean that it's not there; there are muscle groups that are felt more easily than others in the various exercises in which they are used.

LEVEL	REPS	DURATION
BEGINNER	2	20 sec
INTERMEDIATE	2	25 sec
ADVANCED	2	30 sec

CAUTION

This exercise involves no significant risks, but it may be appropriate to kneel on a padded surface, such as a mat, to avoid hurting your knees.

INDICATION

For track and road racers, especially time racers, because of their exaggerated forward lean on the bicycle, and even more particularly if they don't have good flexibility in the gluteus muscles and the ischiotibials.

Seated Upper Body Bend with Pull

START
Sit on the ground with both legs stretched out in front nearly completely. You will touch the ground with your gluteus muscles and your heels. Place your hands behind your knees and grip firmly. Hold your back perpendicular to the ground and look straight ahead.

TECHNIQUE
Bend your upper body so that your face moves closer to your knees. Pull with your arms to increase the pull and produce the best possible stretch in the spinal erector muscles. Avoid bending your knees to move them toward your face, because that will not add to the stretch.

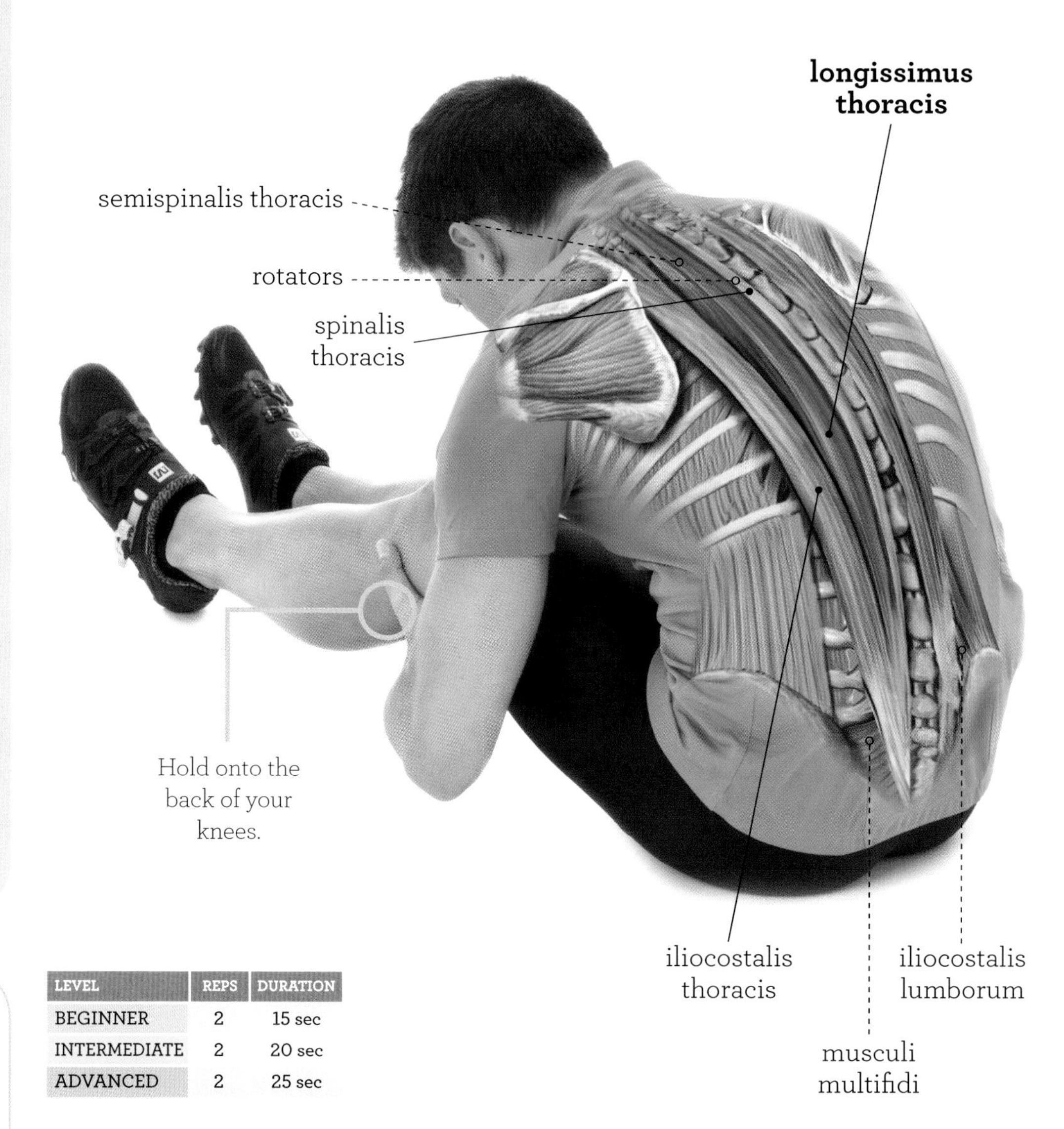

Starting Position

LEVEL	REPS	DURATION
BEGINNER	2	15 sec
INTERMEDIATE	2	20 sec
ADVANCED	2	25 sec

CAUTION
If you experience discomfort in the lumbar area, whether chronic or while doing this exercise, reduce the intensity or stop doing it.

INDICATION
For cyclists whose sports require a very aerodynamic lean on the bicycle, such as track and road riders, especially time racers, due to the tension to which the paravertebral muscles are subjected. Also for regular practitioners of spinning.

Seated Upper Body Bend

START
Sit on the ground with your legs together and your knees straight, so that you are supported on your gluteus muscles, your legs, and your heels. Keep your spine perpendicular to the ground and your hands on either side of your body with your palms resting on the ground and slightly ahead of your upper body.

TECHNIQUE
Slide both hands forward toward your feet, while bending your upper body as necessary to produce tension in the spinal erector muscles. Do the exercise slowly and gradually, and back up if you feel any discomfort in the lumbar area.

LEVEL	REPS	DURATION
BEGINNER	2	20 sec
INTERMEDIATE	2	25 sec
ADVANCED	2	30 sec

CAUTION

Concentrate on the spinal flex, not on anteversion of the pelvis, because that would prioritize the stretch in the ischiotibial muscles rather than the paravertebrals.

INDICATION

For all cyclists who spend a lot of their time in a very aerodynamic lean on the bicycle, because of the strain that this involves for the paravertebral muscles, especially if the cyclist does not have very good flexibility in the hip extensor muscles.

Lateral Spine Tilt

START
Stand next to your bicycle or some other support of a similar height, and hold the forward end of the saddle with one hand. Place your free hand onto the back of your neck and keep your feet far enough apart to ensure stability while you perform the exercise.

TECHNIQUE
Bend your upper body sideways toward the bicycle to produce a unilateral stretch in your paravertebral muscles. As your head moves closer to the bicycle frame, your hold on the saddle will become increasingly useful for support, and you will be able to finish with your forearm resting lengthwise on the saddle.

LEVEL	REPS	DURATION
BEGINNER	2	25 sec
INTERMEDIATE	2	30 sec
ADVANCED	3	30 sec

Starting Position

CAUTION
Start from a stable position and use the bicycle or some other support of similar height as a third support point.

INDICATION
For track and road riders, especially if their sport requires a very aerodynamic position on the bicycle, such as pursuit cycle and time racers, and even more particularly if they have powerful ischiotibials and gluteus muscles with limited flexibility.

Squat with Upper Body Bend

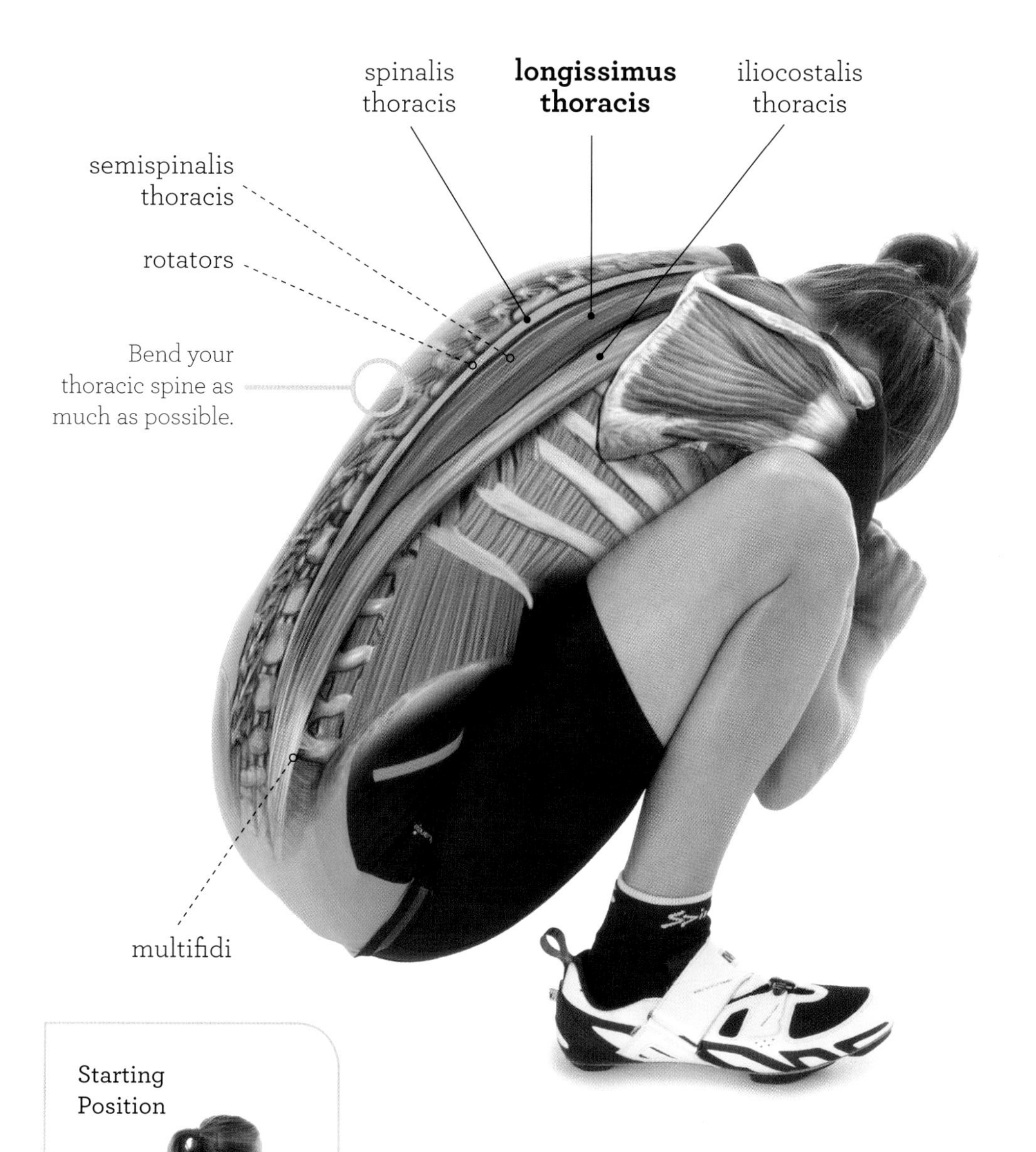

START

Squat with your feet and knees a short distance apart to produce a more stable position. Your arms are between your knees and your elbows are bent so that your hands are in front of your chin. Your spine should remain straight until you begin the stretch.

TECHNIQUE

Lower your head as you bend your neck and upper body. Move your elbows downward as far as you can. You will clearly feel the tension in the paravertebral muscles in the middle and top of your back, which will indicate the effectiveness of the exercise.

LEVEL	REPS	DURATION
BEGINNER	2	25 sec
INTERMEDIATE	2	30 sec
ADVANCED	3	30 sec

Starting Position

CAUTION

It is advisable to keep your feet far enough apart to reduce the instability that this crouch commonly involves.

INDICATION

For all cyclists, especially ones who use very aerodynamic positions on the bicycle. Also for cyclists who lack good flexibility in the gluteus muscles and the ischiotibials, because this forces them to curve their spine more.

Seated Side Bend

START
Sit down and lean to one side while placing the hand on that side on the ground to act as a support. The sole of the foot opposite the support hand touches the ground, in front of your body, so that your knee is bent about 90°. The other leg crosses under it so that its outside is in contact with the ground.

TECHNIQUE
Slowly bend the elbow of your support arm so that your upper body moves to the side and you remain seated on the side of your gluteus. At the same time, let your back bend to the side under the weight of your upper body, and bend your lumbar spine slightly to produce the greatest degree of stretch.

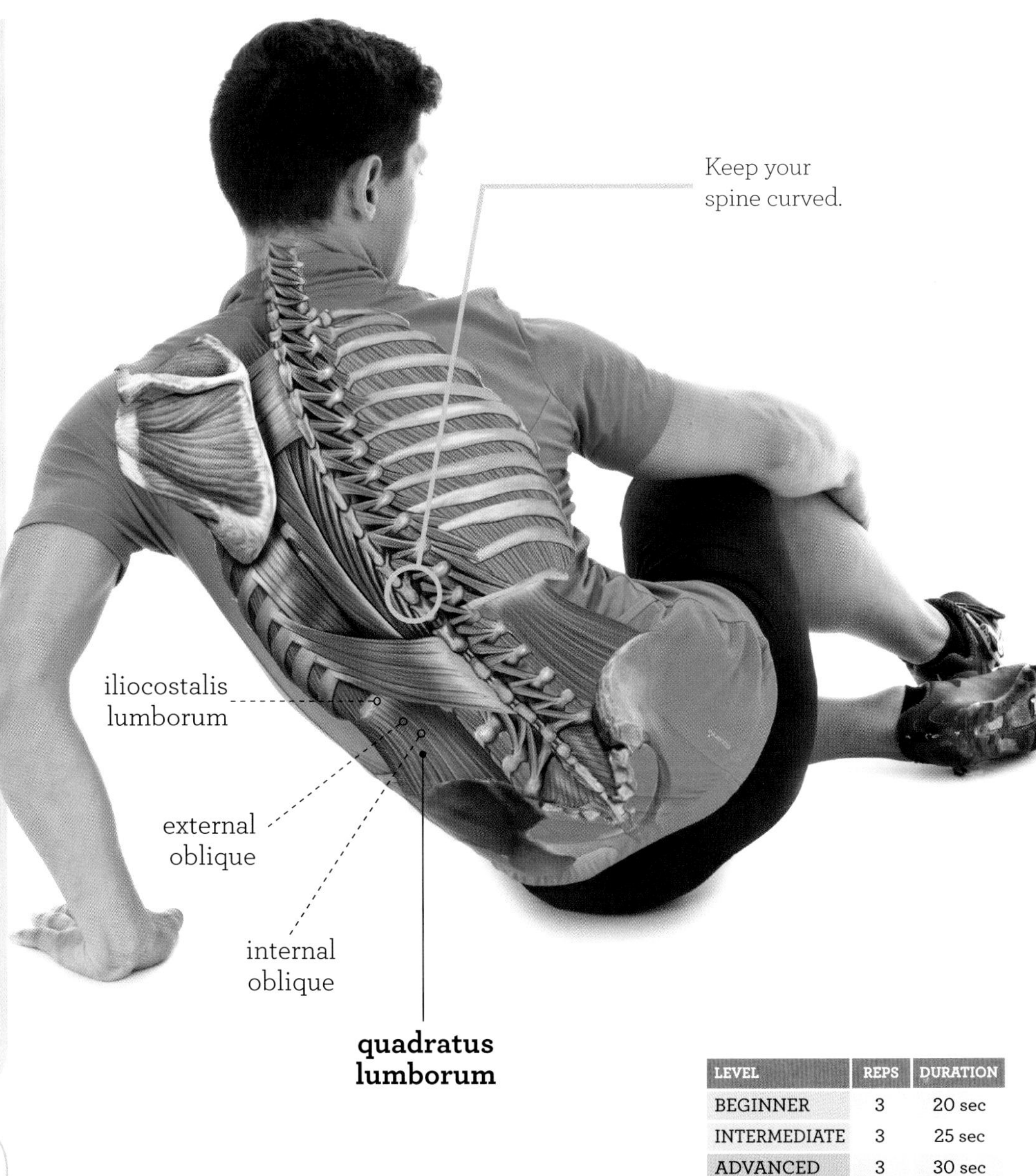

LEVEL	REPS	DURATION
BEGINNER	3	20 sec
INTERMEDIATE	3	25 sec
ADVANCED	3	30 sec

Starting Position

CAUTION
This stretch involves no specific risk, but it's a good idea to perform it on a mat.

INDICATION
For all cyclists, especially ones who experience lumbar pain due to muscle strain in this area, which may be due to poor pedaling technique, improper saddle height, and even excessive, sustained retroversion of the pelvis while pedaling.

Upper Body Bend with Pull

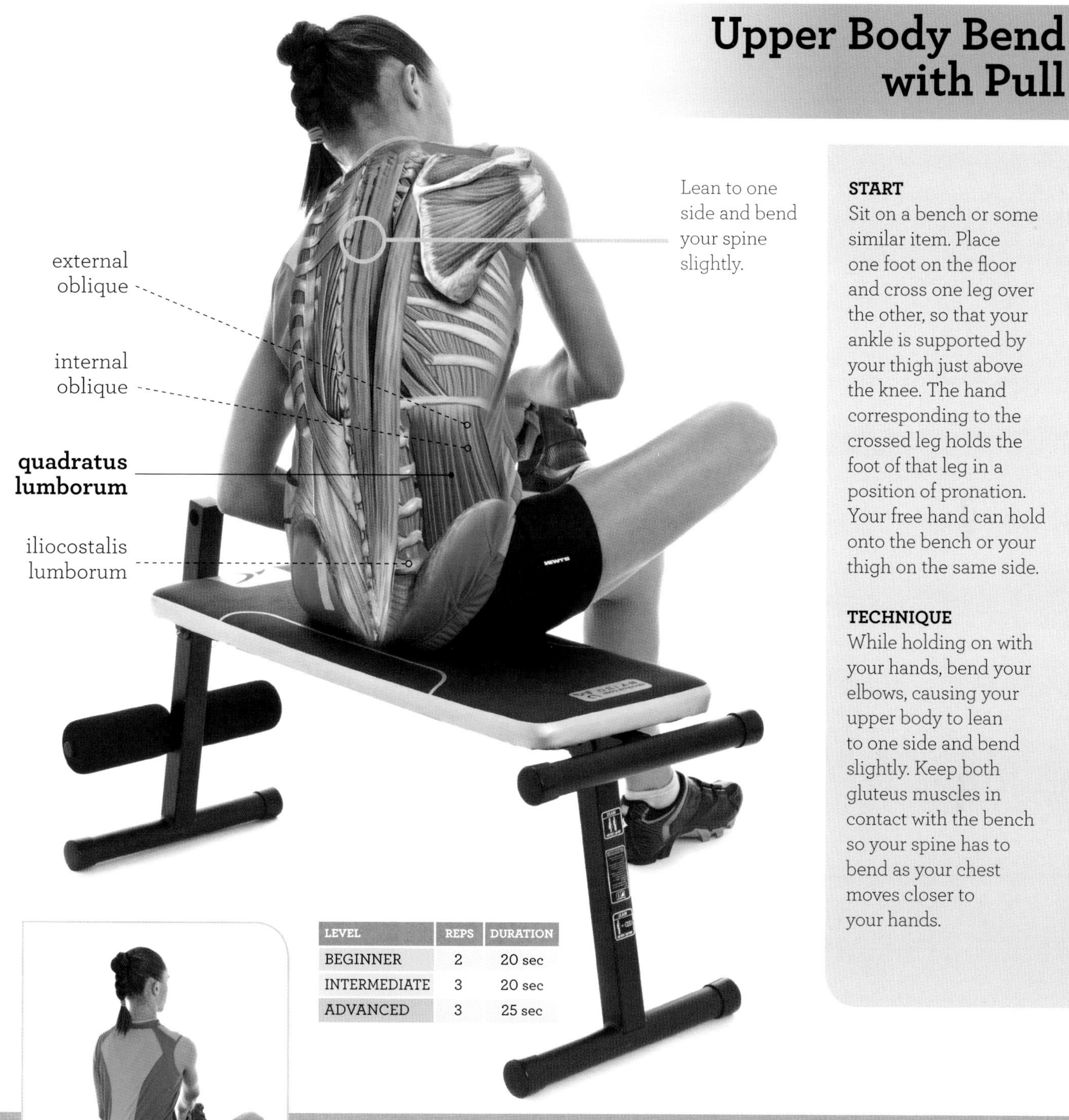

START
Sit on a bench or some similar item. Place one foot on the floor and cross one leg over the other, so that your ankle is supported by your thigh just above the knee. The hand corresponding to the crossed leg holds the foot of that leg in a position of pronation. Your free hand can hold onto the bench or your thigh on the same side.

TECHNIQUE
While holding on with your hands, bend your elbows, causing your upper body to lean to one side and bend slightly. Keep both gluteus muscles in contact with the bench so your spine has to bend as your chest moves closer to your hands.

LEVEL	REPS	DURATION
BEGINNER	2	20 sec
INTERMEDIATE	3	20 sec
ADVANCED	3	25 sec

CAUTION
Be particularly careful if you experience lumbar pain, and reduce the intensity of the exercise if you feel any discomfort.

INDICATION
For cyclists who experience muscle pain in the lumbar area due to strain (to which riders who rock their hips excessively while pedaling are more liable), for cyclists whose position is very curved, and for cyclists whose bicycles are not adjusted to their physical dimensions.

Kneeling Side Bend

START
Kneel down and move your weight rearward so that the backs of your thighs touch your calves. Your ankles will be in plantar flexion, with your instep touching the ground. Bend your upper body forward with your hands resting on the ground, and slide them forward so that they are together or next to one another.

TECHNIQUE
Slide your hands to one side along the ground without moving your knee supports so that your back arches and your lumbar vertebrae bend slightly. Move your hands as far as possible without changing your leg supports, and keep your hips in the same position.

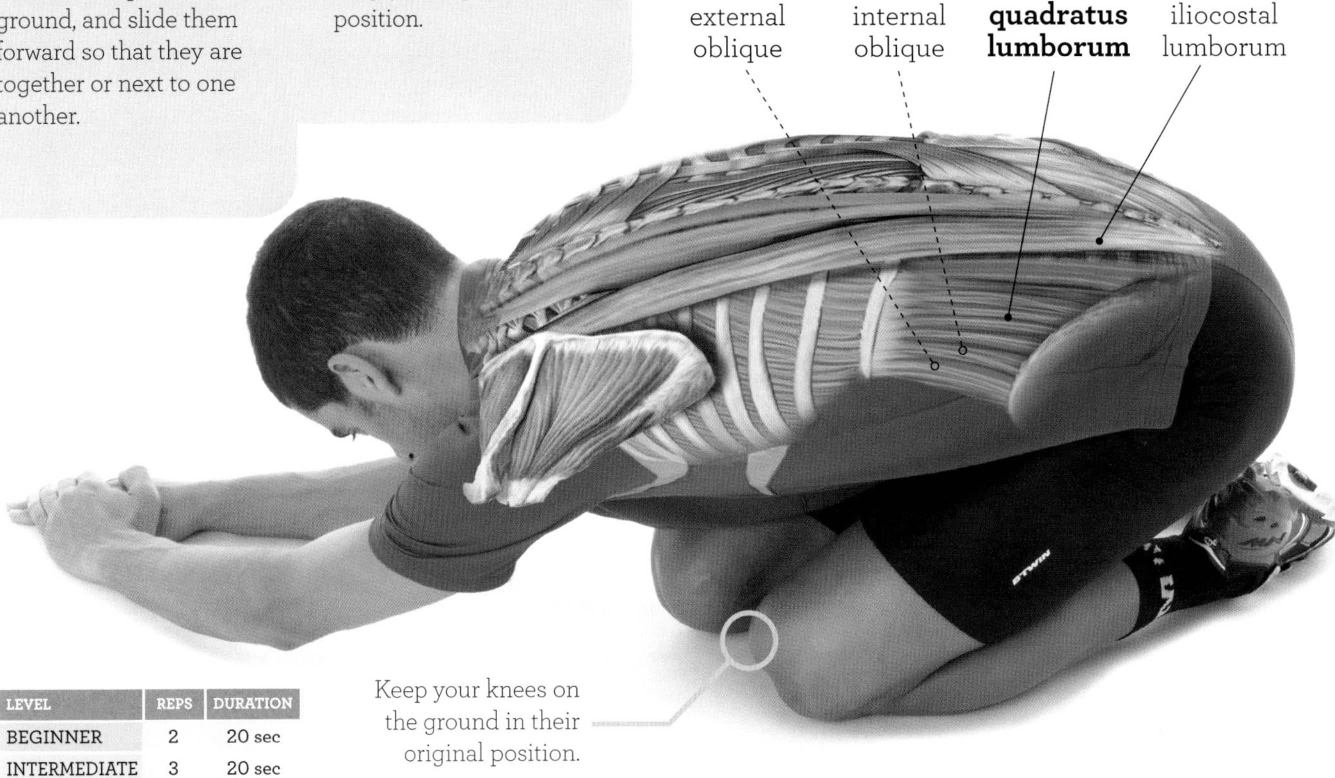

LEVEL	REPS	DURATION
BEGINNER	2	20 sec
INTERMEDIATE	3	20 sec
ADVANCED	3	25 sec

CAUTION

Do this exercise on a mat or a slightly padded surface to avoid pain in your support points, especially your knees and ankles.

INDICATION

For cyclists who experience strain in the lumbar muscles, which is common in riders who maintain a very aerodynamic position on the bicycle, such as track riders, time racers, and others who likewise suffer from strain in this area, because of muscle imbalances, improper pedaling technique, intense training, and other details related to the bicycle.

Cobra Position

LEVEL	REPS	DURATION
BEGINNER	2	15 sec
INTERMEDIATE	2	20 sec
ADVANCED	2	25 sec

START

Lie face down and place your palms on the ground on each side of your chest as if you were about to do a push-up. Your hips must be in contact with the ground, along with your chest and the front of your thighs. Look straight ahead, and keep your abdomen and lower limbs relaxed.

TECHNIQUE

Straighten your arms as if your were trying to get up or to separate your chest from the floor, while keeping your upper body relaxed and your hips in contact with or very close to their original support. Extend your spine as much as possible so that you subject the flexor muscles of your upper body to tension for the time appropriate to your level.

CAUTION

Remember to keep the muscles of your upper body and your legs relaxed in order to facilitate the stretch, and to use a mat whenever possible. Because their functions are so important, it is also a good idea to combine stretches for the abdominal muscles with exercises to strengthen them.

INDICATION

For all cyclists, especially ones who take part in long races, because of the shortened position of the abdominal muscles during hours of pedaling and their reduced activity and participation in this sport

Arch Position

START
Kneel down with your upper body and neck perpendicular to the ground. Your support points are your knees and your toes, such that your ankles are in nearly complete dorsal flexion. Hold your heels firmly with your hands so that they provide support.

TECHNIQUE
Extend your upper body as you move your gluteus muscles away from your calves, so that your chest, abdomen, and thighs form an arc and your head is the rearmost part of your body. At this point, your hands will support a good part of your weight, so your arms and feet will need to be lined up with one another to form a solid support.

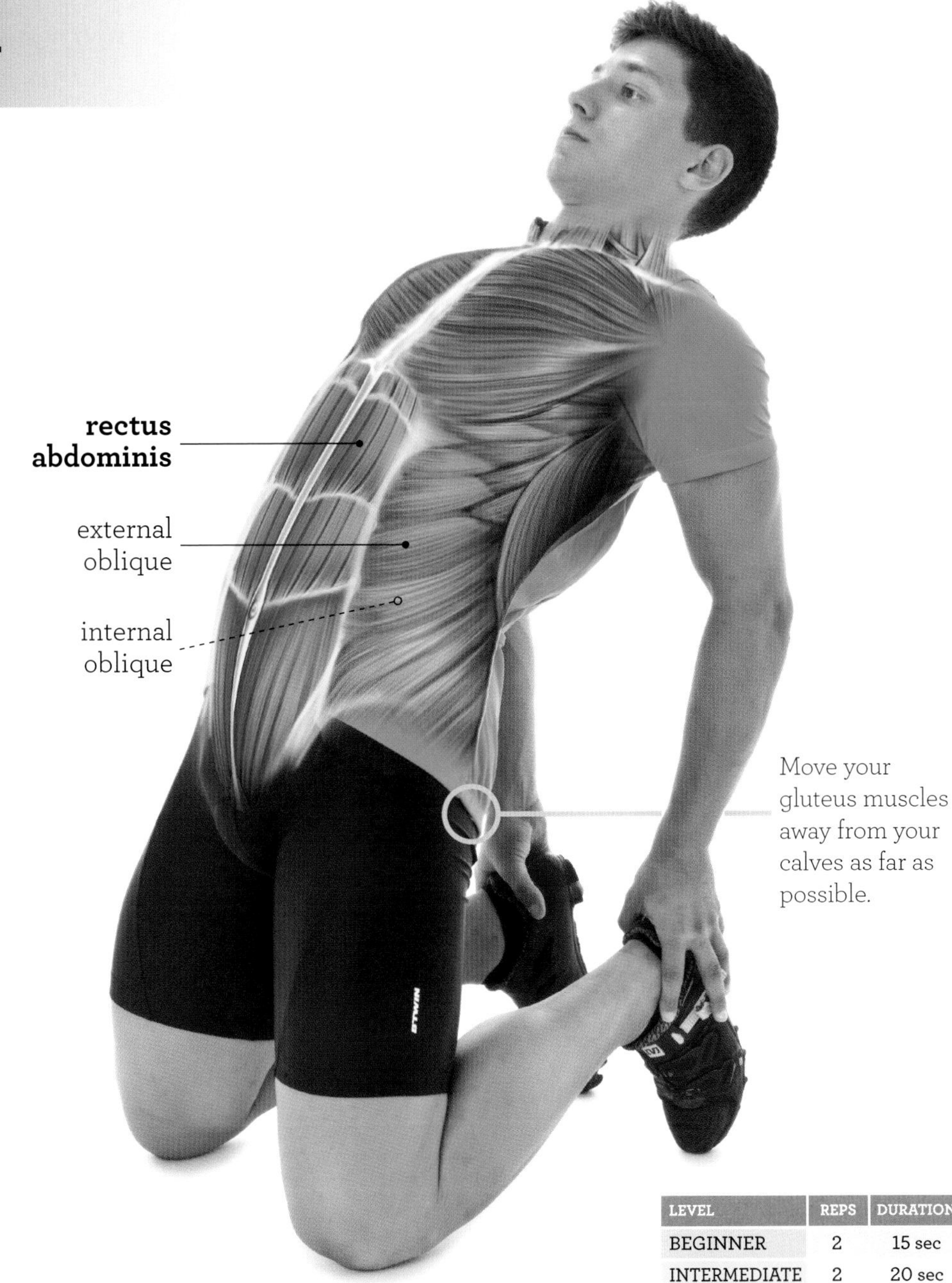

Move your gluteus muscles away from your calves as far as possible.

LEVEL	REPS	DURATION
BEGINNER	2	15 sec
INTERMEDIATE	2	20 sec
ADVANCED	2	25 sec

Starting Position

CAUTION

This exercise involves no risk, but it is not easy to do, so you will have to assess whether it is right for you. Start from a balanced position and use a mat to avoid hurting your knees.

INDICATION

For all cyclists, especially for road and track racers, because of the shortened position of the abdominal muscles and their lack of involvement while pedaling.

Upper Body Rotation

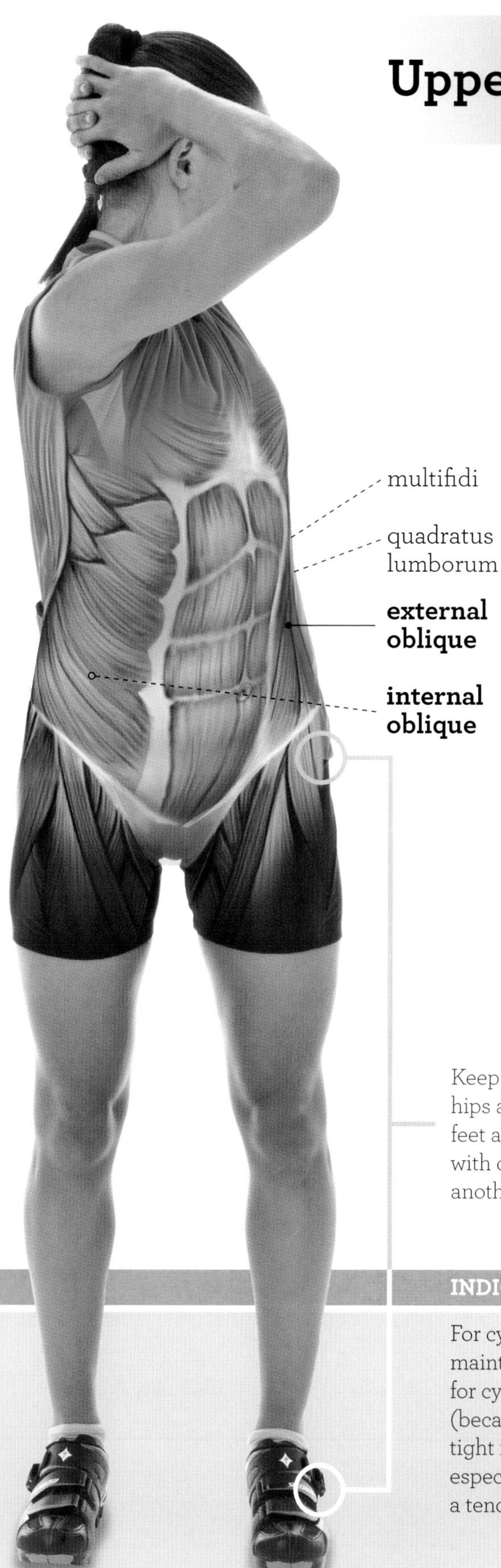

LEVEL	REPS	DURATION
BEGINNER	2	15 sec
INTERMEDIATE	2	20 sec
ADVANCED	2	25 sec

START

Stand with your knees straight, your feet in line with your shoulders, and your upper body perpendicular to the ground. Place your hands behind your neck with your fingers interlaced. Move your elbows outward and relax your abdomen to increase the effectiveness of the exercise.

TECHNIQUE

Rotate your upper body so that your shoulders and hips are no longer in alignment. It is important that your hips and feet remain in alignment during the entire exercise, because the torsion should take place between the hips and the chest, rather than between the chest and the feet. At the end of the movement, you should remain looking to one side while you simultaneously stretch the external and internal obliques on each side.

CAUTION

Avoid moving your hips and legs to maximize the range of movement. This will not improve the stretch, and it could even reduce its effectiveness.

INDICATION

For cyclists whose sports require that they maintain a very aerodynamic position and for cyclists who lean far over their bicycles (because the abdominal muscles remain tight for a long time and are inactive). It is especially appropriate for cyclists who have a tendency to arch their backs.

Upper Body Rotation with Support

START

Stand next to your bicycle, midway between the saddle and the handlebar stem, as if you were walking along with it. Hold the handlebars with both hands and squeeze both brake levers to create a solid support on the bicycle. At this point, your feet, hips, and shoulders should be aligned with one another, or nearly so.

TECHNIQUE

Change the position of your feet, legs, and hips, as if the lower half of your body wanted to turn away from the bicycle. Your hands must remain in their original support position, and so does the bicycle. This will require a rotation of the upper body as far as possible, in order to subject the oblique muscles on both sides to the stretch.

Starting Position

LEVEL	REPS	DURATION
BEGINNER	2	20 sec
INTERMEDIATE	2	25 sec
ADVANCED	3	25 sec

CAUTION

Continue squeezing the brake levers and tilt the bicycle slightly toward you for the best balance while performing this exercise.

INDICATION

For all cyclists, because of the reduced activity of the abdominal muscles while riding, and especially for riders who use a low, curved position on the bicycle, because of the compressed or shortened position of their abdominal muscles.

Supine Upper Body Rotation

START

Lie down on your back on the ground and place your hands behind the back of your neck for comfort, or else place them on the ground on both sides of your body with your arms held straight out (if you want to increase stability). Bend your hips and knees to 90°, as if you were sitting. Your neck should be relaxed and you should look upward.

TECHNIQUE

Rotate your lower body in such a way that the top of your back remains in contact with the ground and the lower part gradually separates from it. Try to lower the outer part of your thigh closest to the ground gradually so that the two come into contact with one another. Your legs should remain together throughout the exercise.

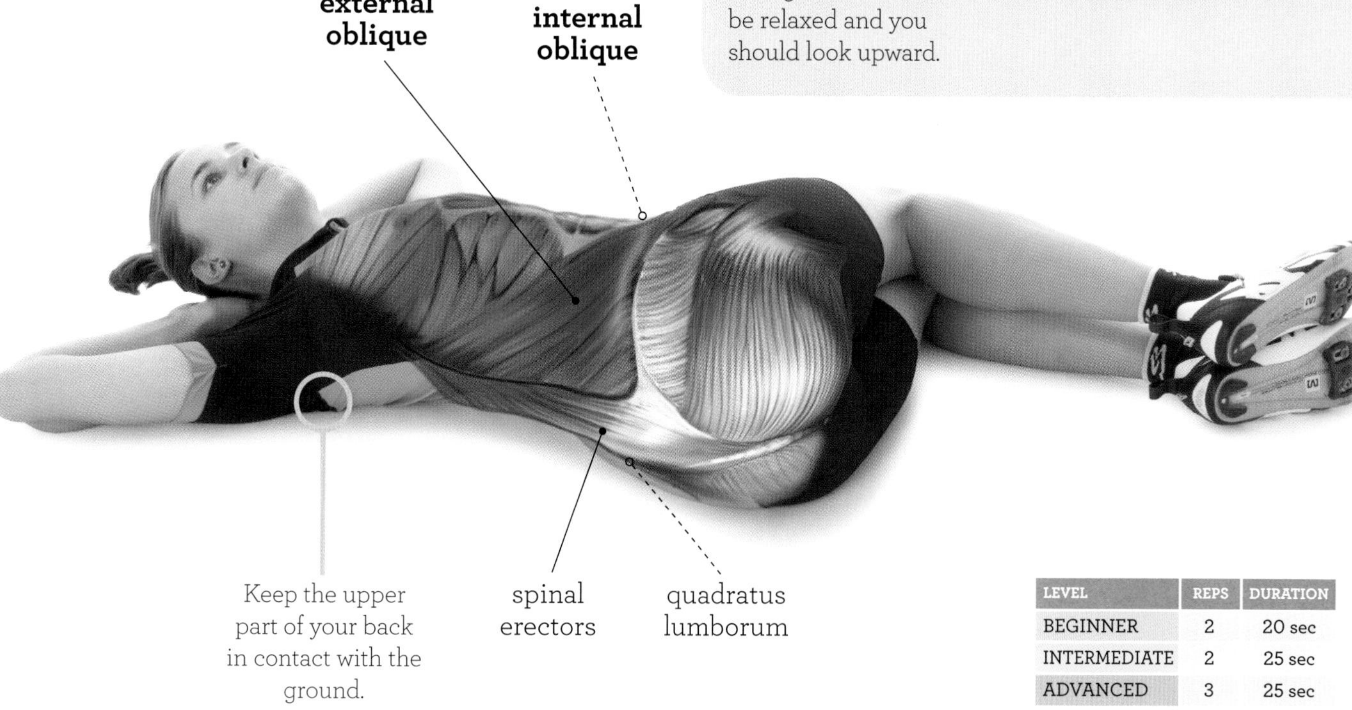

LEVEL	REPS	DURATION
BEGINNER	2	20 sec
INTERMEDIATE	2	25 sec
ADVANCED	3	25 sec

CAUTION

Do not attempt to raise your head, because that would place unnecessary tension on the neck muscles during the exercise.

INDICATION

For track or road riders, particularly for those who use a very aerodynamic position, and for bicyclists who have powerful but not very flexible hip extensor muscles, such as some time racers, sprinters, and speed and pursuit racers, whether as individuals or as part of teams.

BTWIN

SHOULDER AND UPPER LIMB STRECTCHES

SHOULDER STRETCHES

The shoulder is a joint with a much greater range of movement than most other joints in the body. This is an advantage, because of the number of movements it allows, but at the same time it is a disadvantage because of its relative instability. For cyclists, the upper limbs are one of the support elements on the bicycle, because they rest on the handlebars; other support points are the saddle and the pedals. As the links between the upper limbs and the torso, the shoulders must contribute to the stability and control of the handlebars, to constant rebalancing on them, and to absorption and partial dampening of the impacts that result during cycling. Standing on the pedals to climb a steep hill, sprinting, and initiating a breakaway require that cyclists redistribute their weight in order to move their position forward. Add to this the need to move the bicycle from side to side, with that characteristic weaving movement produced mainly by the shoulder muscles. This same tension is experienced by downhill racers, mountain bikers, and motocross (BMX) riders, because these muscles partly absorb the shocks produced by rolling over rough terrain. Thus, in order to avoid muscle imbalances resulting from cycling sports, it is important to reinforce and maintain good flexibility in the muscles that move and stabilize the shoulder.

DELTOID

This muscle has three sections that are easy to identify; they are known by the different functions that each one performs. They all insert at the deltoid tuberosity of the humerus.

Clavicular or anterior portion: This originates at the distal third of the clavicle. Its main function is the flexion or the antepulsion of the shoulder, although it also contributes to its internal rotation and adduction.

Acromial or medial portion: This originates at the acromion, and its function is shoulder abduction.

Spiny or posterior portion: This originates at the spine of the scapula, and its main function is the extension or retropulsion of the shoulder, although it also contributes to its external rotation and adduction.

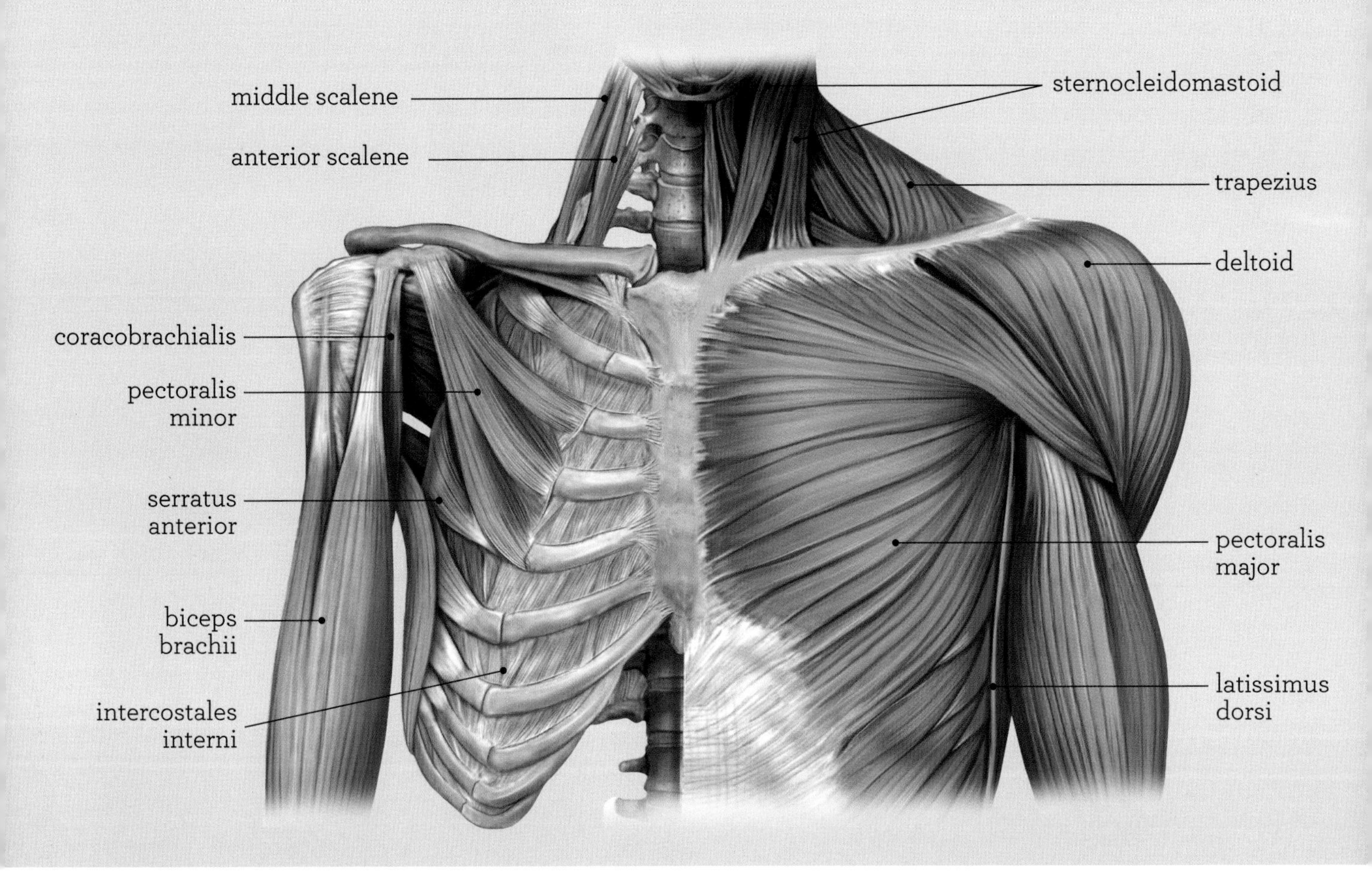

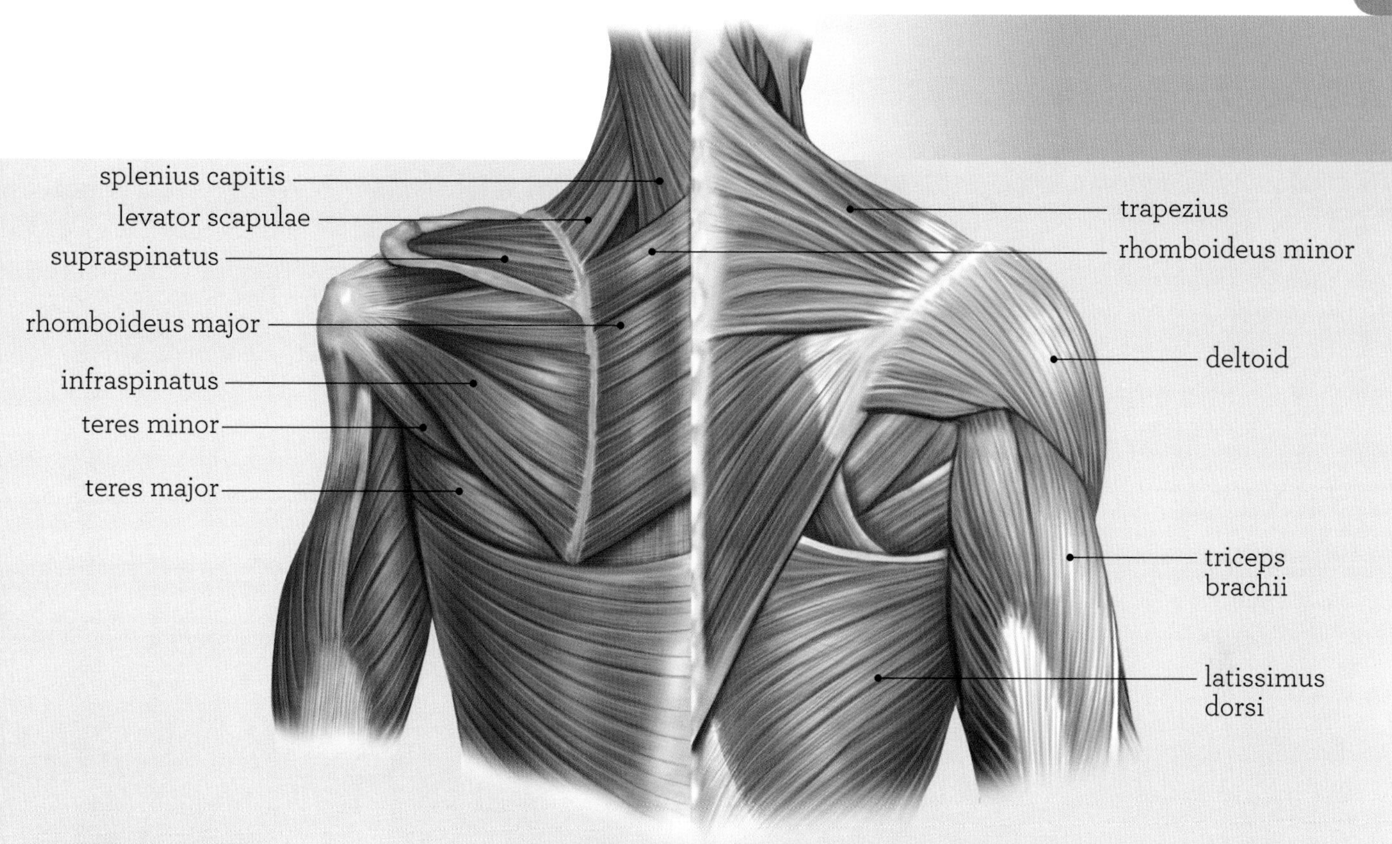

PECTORALIS MAJOR
Like the deltoid, this powerful muscle has three different sections, but because of their arrangement, their functions are not so different. The three sections share an insertion located on the crest of the tuberculum majus of the humerus.
Clavicular or upper portion: This originates at the middle half of the clavicle; its main functions are adduction and flexion or antepulsion of the shoulder.
Sternocostal or middle portion: This originates at the sternum and the costal cartilage of ribs one through seven. Its main function is shoulder adduction, especially horizontally, although it contributes slightly to retropulsion.
Abdominal or lower portion: This originates at the aponeurosis of the external oblique of the abdomen. Its main function is shoulder adduction, but it also lowers the arms from a raised position.

It is important to point out that, even though all these portions perform shoulder adduction, their involvement increases and decreases depending on the angle of flex.

CORACOBRACHIALIS
This muscle originates at the coracoid apophysis of the scapula, and it inserts at the middle third of the humerus. Its main function is antepulsion or flexion of the shoulder, but it plays a small role in the shoulder's adduction and internal rotation.

LATISSIMUS DORSI
This muscle originates at the spiny apophyses of vertebrae T7 to L5, the posterior part of the sacral bone, and the iliac crest. Its insertion is at the crest of the tuberculum minus of the humerus. This powerful muscle's main function is shoulder adduction, although it is also involved in the shoulder's internal rotation and retropulsion.

TERES MAJOR
This muscle originates at the lower angle of the scapula, and it inserts at the tuberculum minus of the humerus. Its main function is internal rotation of the shoulder, and it participates to a lesser degree in the shoulder's adduction and retropulsion.

TERES MINOR
This muscle originates at the lateral edge of the scapula, and it inserts at the tuberculum majus of the humerus. Its main function is external rotation of the shoulder.

SUPRASPINATUS
This muscle originates at the fossa supraspinata of the scapula, and it inserts at the tuberculum majus of the humerus. Its function is abduction of the shoulder, although it contributes moderately to this movement.

INFRASPINATUS
This muscle originates at the fossa infraspinata of the scapula, and it inserts at the tuberculum majus of the humerus. Its function is external rotation of the shoulder.

Bilateral Retropulsion

START

Stand with your feet about shoulder width apart. Your upper body and neck should be perpendicular to the ground and remain that way while performing the exercise. Place your hands behind your back and hold them together just below the wrists, as if you were waiting for something.

TECHNIQUE

Keeping your hands together, try to raise them upward behind your body by means of shoulder retropulsion. Try to achieve the maximum range of movement and hold it without bending your upper body forward. It will take considerable effort to hold the position, because this is an active stretch, so you will need to pay particular attention to holding it, especially during the final seconds of the repetition.

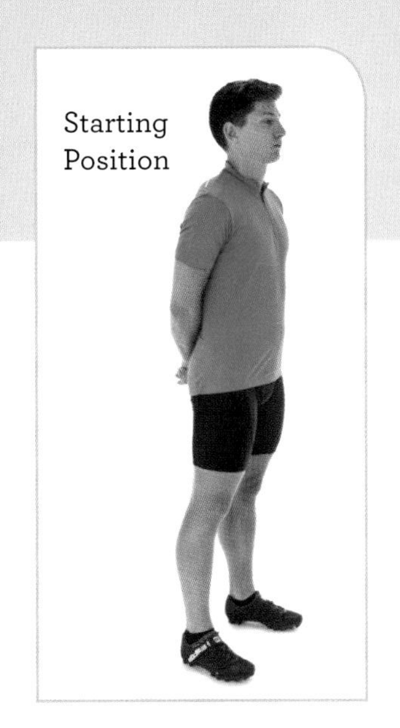
Starting Position

deltoid

coracobrachialis

pectoralis major

Keep your upper body perpendicular to the ground.

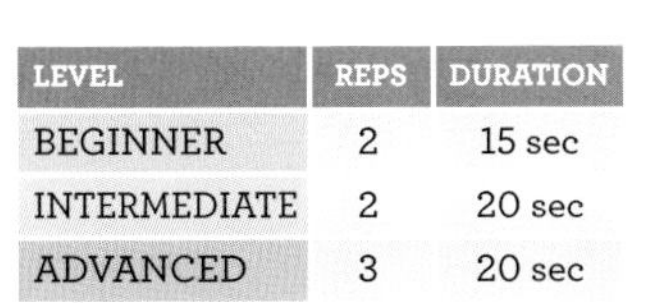

LEVEL	REPS	DURATION
BEGINNER	2	15 sec
INTERMEDIATE	2	20 sec
ADVANCED	3	20 sec

CAUTION

There is a tendency to lean the upper body forward, because this gives the sensation of stretching farther; however, this is a deceptive sensation, so try to keep your upper body perpendicular to the ground, even though you cannot raise your hands as far.

INDICATION

For all cyclists, especially those whose sports require spending a long time on the bicycle, such as in long-stage races and events over rough terrain that entail constant shock absorption and vibration of the handlebars.

Rearward Pull with Anchor Point

START
Stand with your back toward the bike, holding the saddle with one hand and the handlebar with the other. You can also use a handrail of the proper height, or wall bars. Your knees and elbows must be straight, with your upper body perpendicular to the ground. Remember to place your feet at least 8 inches (20 cm) from the bike. This will allow you to move without interference from the bicycle.

TECHNIQUE
While keeping your hands in place, pivot 90° so that you are facing the same direction as the bike. You probably will feel tension in the front and middle part of your deltoid. If so, hold the position for a few seconds. If you do not feel any tension, you can turn farther than 90° or lower your center of gravity by bending your knees slightly.

LEVEL	REPS	DURATION
BEGINNER	2	15 sec
INTERMEDIATE	2	20 sec
ADVANCED	3	20 sec

Starting Position

CAUTION

Keep pressure on the brake lever with the hand on the handlebar, and keep your feet far enough from the bicycle for proper, risk-free performance of the exercise.

INDICATION

For all cyclists, especially if they experience discomfort in the shoulders or the upper limbs while riding. Also for athletes who ride a lot or commonly ride over rough surfaces.

Retropulsion with Support

START
Stand facing away from the bicycle and hold onto it with both hands. One hand should grip the saddle and the other, the handlebar stem, so that your palms face rearward. When you have a solid grip, take a short step forward so that your body moves away from the bike frame. Your elbows and knees will be extended, with your feet about shoulder width apart, and your upper body will be perpendicular to the ground.

TECHNIQUE
Slowly bend your knees and hips, as if you were going to sit down. As your center of gravity lowers, you will have to bend your elbows in order to maintain your grip. This will cause a retropulsion of both shoulders and the corresponding stretch in the deltoid, especially in the front part. Avoid leaning your upper body forward too much, because that would reduce the effectiveness and the intensity of the stretch.

deltoid

coracobrachialis

pectoralis major

LEVEL	REPS	DURATION
BEGINNER	2	20 sec
INTERMEDIATE	3	20 sec
ADVANCED	3	25 sec

Starting Position

CAUTION
Keep sufficient distance away from the bike before starting to do this stretch, and lower yourself gradually, because the exercise requires a solid starting position and a certain amount of control in the movements.

Keep some distance away from the bike for greater stability.

INDICATION
For cyclists with discomfort or muscle tension in the shoulders and the upper limbs, especially if they do long workouts, ride over rocky or rough terrain, or perform jumps or acrobatics.

Posterior with Arms in Front

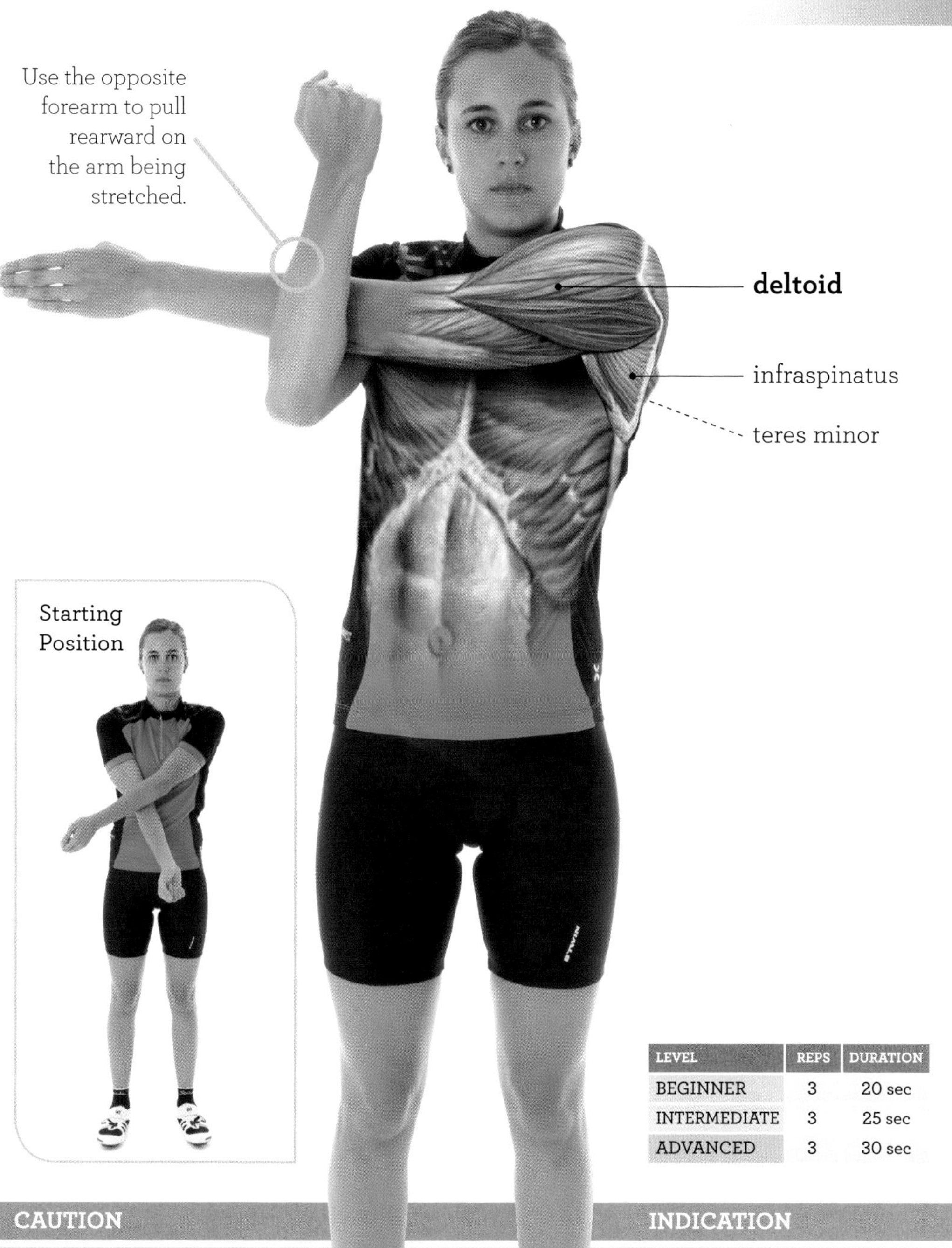

START
Hold one arm in front of your body, with your elbow straight and your palm facing forward. Then cross the other arm in front of the first one so that the latter is supported in the front of the elbow. Both elbows should be straight, with your hands relaxed. Your back should remain straight, perpendicular to the ground.

TECHNIQUE
Bend the elbow of the lower arm so that the other one is supported and held in place. Pull the supported arm toward you. This will produce maximum frontal adduction and tension in the rear part of the deltoid. When you reach the optimum point of the stretch, hold the pull on the arm for a few seconds before changing arms.

LEVEL	REPS	DURATION
BEGINNER	3	20 sec
INTERMEDIATE	3	25 sec
ADVANCED	3	30 sec

CAUTION

This exercise involves no risk, given its simplicity and the control of the intensity at all times.

INDICATION

For maintaining optimum overall flexibility in cycling and other sports. Even though the spiny part of the deltoid does little work while riding, it's a good idea to stretch it.

Self-assisted Unilateral Adduction

START
Stand with your upper body and neck perpendicular to the ground, and maintain this position while performing the stretch. Place both arms behind your back, with one hand firmly gripping the opposing wrist, as if you were waiting for something.

TECHNIQUE
Pull on your wrist with the hand holding it. This will cause maximum adduction of the shoulder behind your back. Keep your upper body perpendicular to the ground, and you will feel the tension in the clavicular and acromial parts of the deltoid as the movement approaches its limit.

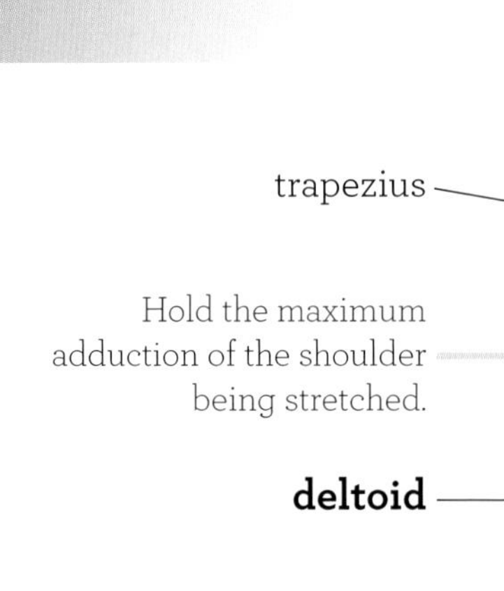

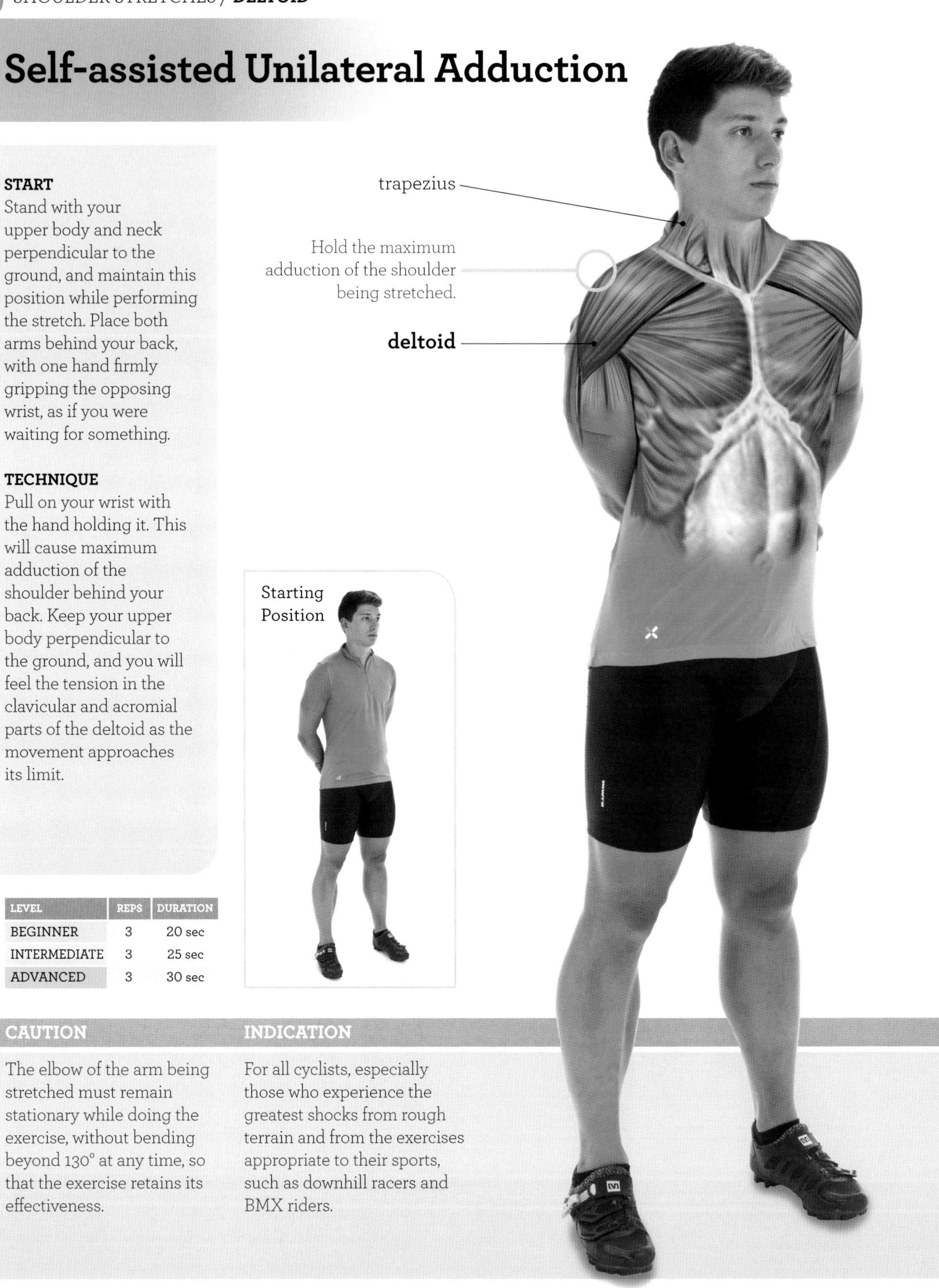

Starting Position

LEVEL	REPS	DURATION
BEGINNER	3	20 sec
INTERMEDIATE	3	25 sec
ADVANCED	3	30 sec

CAUTION

The elbow of the arm being stretched must remain stationary while doing the exercise, without bending beyond 130° at any time, so that the exercise retains its effectiveness.

INDICATION

For all cyclists, especially those who experience the greatest shocks from rough terrain and from the exercises appropriate to their sports, such as downhill racers and BMX riders.

Support with Bent Elbow

START

Stand next to a vertical support, such as a post or a door frame. Raise the arm closest to the support and bend your elbow so that your arm forms a 90° angle with your body and at your elbow. Your palm will face forward with your fingers pointing upward. Without changing your arm position, rest the front of your forearm and your palm on the support.

TECHNIQUE

Try to pivot your upper body toward the side opposite the support without moving your feet. Your support arm should remain immobile, acting as a block and remaining behind your upper body. The farther you move, the clearer the tension in the pectoralis major and the front part of the deltoid will become. Hold the position of maximum stretch before changing arms.

Starting Position

LEVEL	REPS	DURATION
BEGINNER	3	20 sec
INTERMEDIATE	3	25 sec
ADVANCED	3	30 sec

CAUTION

You can place your forearm on a slightly higher support if that helps you feel the stretch, but do not use a lower support, because that would limit the range of movement. If you feel any joint pain in the shoulder, reduce the intensity of the stretch.

INDICATION

For all cyclists, because of the muscle imbalances they may experience between the front and rear parts of the upper body, due to the nature of the riding position. These could lead to posture problems.

Unilateral Vertical Support

START
Stand next to a solid, vertical support that will not move in response to the pull you will exert. This could be a wall, a column, or a post. The support should be at your side, and your near foot will be on the ground behind the other one. Hold onto the vertical support at about shoulder height or slightly higher with the hand on the side being stretched, and keep your elbow slightly bent.

TECHNIQUE
Step forward with the rear foot so that you step beyond the support. Keep your hand in place without rotating your upper body, and look straight ahead in order to force the shoulder extension and the stretch in the pectoralis major and the front of the deltoid. Remember that the elbow of the support arm should remain slightly bent to keep from transferring the stretch to the corresponding flexor muscles.

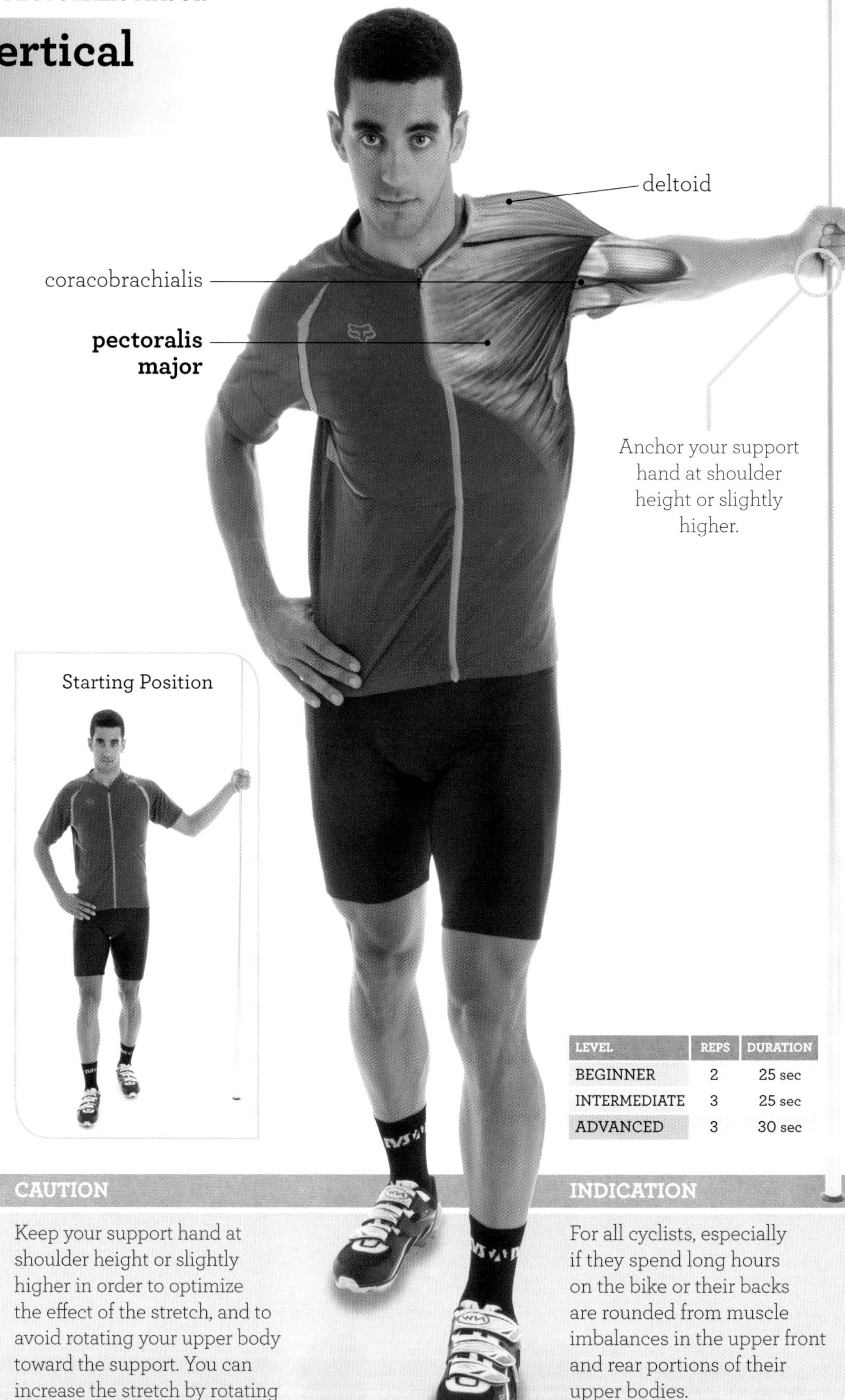

LEVEL	REPS	DURATION
BEGINNER	2	25 sec
INTERMEDIATE	3	25 sec
ADVANCED	3	30 sec

CAUTION
Keep your support hand at shoulder height or slightly higher in order to optimize the effect of the stretch, and to avoid rotating your upper body toward the support. You can increase the stretch by rotating your upper body away from the support.

INDICATION
For all cyclists, especially if they spend long hours on the bike or their backs are rounded from muscle imbalances in the upper front and rear portions of their upper bodies.

Seated Pull with Assist

Rest your back firmly against your partner's thigh and hip.

LEVEL	REPS	DURATION
BEGINNER	2	20 sec
INTERMEDIATE	2	25 sec
ADVANCED	2	30 sec

START

You will need the help of an assistant for this exercise. Sit down and fold your hands behind your neck. Your assistant will take a position behind you, kneeling on one knee and touching your back, so that his or her thigh and hip act as a backing and a support when you perform the stretch. Your assistant will hold you by the front of your elbows with palms facing rearward.

TECHNIQUE

Your assistant will pull back on your elbows while supporting your back with his or her thigh and hip. The movement should be slow and controlled, and you will need to keep your hands on the back of your neck without letting go.

As the horizontal abduction of your shoulder progresses, you will clearly feel tension in your pectoral muscles, and the stretch will become more effective.

CAUTION

In stretches involving the help of an assistant, it is important to proceed very slowly with consistent communication between the two people, in order to stay within the safety limits of the exercise and avoid possible injury.

INDICATION

For all cyclists, given the position on the bicycle, especially on racing bikes, and also for cyclists with muscle imbalances in the upper part of the torso or with rounded backs, as well as for cyclists in sports involving jumps, acrobatics, or rough surfaces that may produce tension in the pectoral muscles.

Horizontal Abduction with Support

START

Stand beside the saddle of your bicycle. Align your feet with your hips and lean your upper body forward. Hold the tip of the saddle with the hand closest to the bicycle and rest your forearm on the saddle. Grip the handlebar with your other hand. You can rest your forearm on the handlebar to reduce pressure on your lumbar vertebrae.

TECHNIQUE

Move the leg closest to the bicycle rearward and bend the knee of your other leg so that your chest moves downward and you produce horizontal abduction in the shoulder being stretched. Your body weight will increase this horizontal abduction, so it will not be necessary to add more force to produce an effective stretch in the pectoralis major and the other muscles.

LEVEL	REPS	DURATION
BEGINNER	2	15 sec
INTERMEDIATE	2	20 sec
ADVANCED	2	20 sec

CAUTION

Before starting this exercise, you should double-check your stability, because you will be resting your hands on a movable object, namely the bicycle. There are more effective exercises for stretching the pectoralis major, so it's appropriate to save this one for times when you experience tension in the pectoral muscles during your athletic activity.

INDICATION

For cyclists who experience tension in their pectoral muscles due to their sports, such as mountain biking, BMX, and downhill racing, or who have muscle imbalances in their upper torsos, with tightness of the muscles in the front.

Shoulder Antepulsion

Pull on your elbow to extend the range.

deltoid

teres major

latissimus dorsi

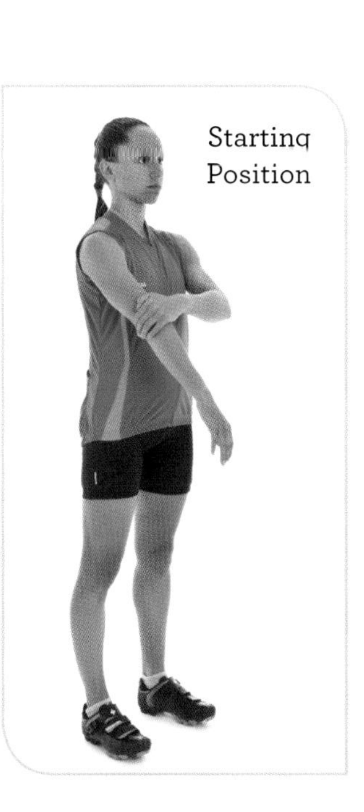
Starting Position

START
Stand with your upper body perpendicular to the ground and look straight ahead. Advance one arm slightly, to a 45° angle with your upper body. Keep your elbow straight and your hand relaxed and in pronation. The opposite hand should hold the outer part of your elbow firmly, but without pulling on it.

TECHNIQUE
Raise your straight arm by means of a flexion or antepulsion of the shoulder. When you reach the greatest degree of flexion, try to extend the movement with the help of the opposite arm. Try to hold the arm being stretched behind your head. If the flexibility of your latissimus dorsi muscle is not very good, you may not reach this point, but this does not mean that you are not stretching properly.

LEVEL	REPS	DURATION
BEGINNER	1	20 sec
INTERMEDIATE	2	20 sec
ADVANCED	2	25 sec

CAUTION

This exercise requires extreme antepulsion of the shoulder, which is a relatively unstable joint; therefore, you should avoid carrying the exercise beyond the point of prudence, and reduce the intensity or stop if you feel any discomfort.

INDICATION

For triathletes, because of the muscle's involvement in swimming. The latissimus dorsi is not a muscle that is used much in cycling, but it should be stretched in the interest of proper overall flexibility.

Bilateral Pull with Support

START
Stand next to your bicycle, facing the frame between the two wheels. Hold onto the top tube of the frame, with both hands together and palms down. Your upper body should be perpendicular to the ground, with your legs straight. You can use some other horizontal support besides the bicycle, as long as it is of a similar or greater height.

TECHNIQUE
Without moving the bicycle, move your feet rearward while bending your upper body forward and producing antepulsion or flexion in both shoulders. Your chest should move downward, and your hand grip will remain firm, with your elbows nearly straight. Your body weight will stretch the latissimus dorsi, so it is not necessary to exert much additional effort.

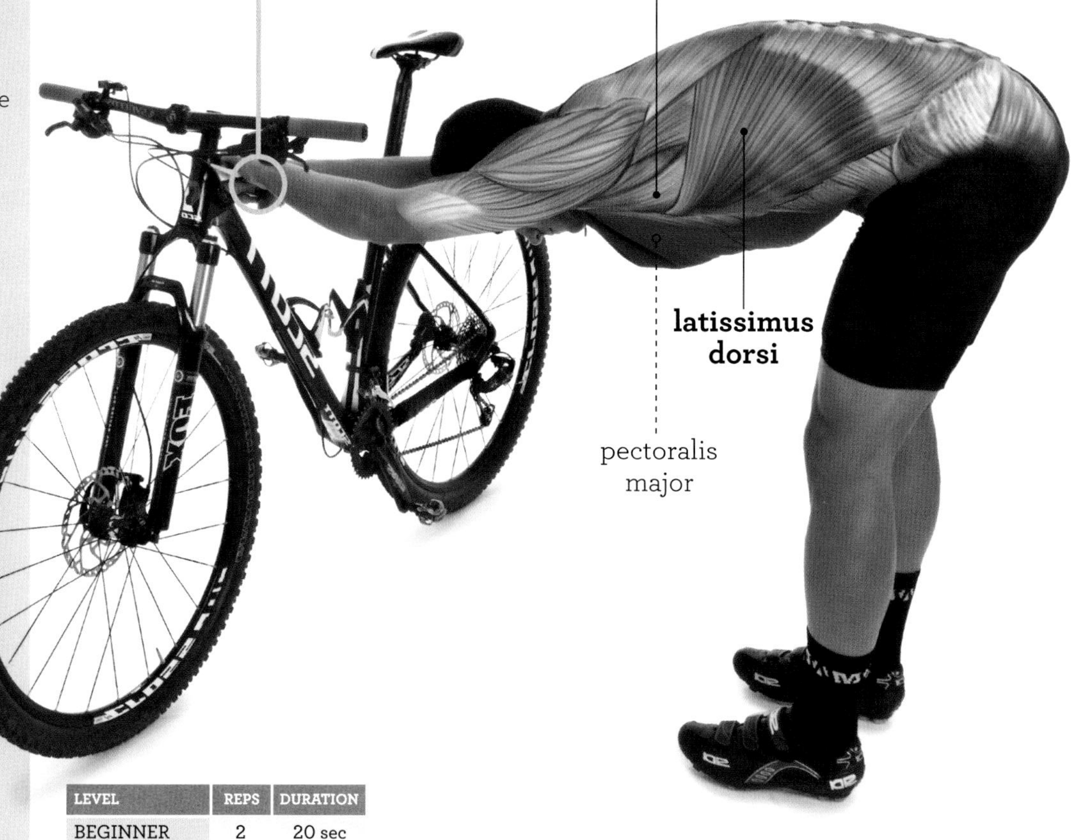

LEVEL	REPS	DURATION
BEGINNER	2	20 sec
INTERMEDIATE	2	25 sec
ADVANCED	2	25 sec

CAUTION

Do this exercised in a slow, controlled manner, and stop if you feel any discomfort in your shoulders, because the shoulder is a relatively unstable joint and it is subjected to maximum antepulsion.

INDICATION

For all athletes who want to maintain good overall flexibility, not just for cyclists; however, it is especially appropriate for athletes who combine cycling with swimming, such as triathletes.

Unilateral Pull with Support

latissimus dorsi

teres major

Hold the middle of the handlebars with your hand palm down.

START
Stand near to the bicycle, but some distance away from it, and face the frame. You should stand even with the rear wheel and the foot closest to the handlebar slightly advanced. Hold the middle of the handlebar with one hand, keep your elbow straight, and lean slightly to the side, with your upper body to maintain the distance from the rear wheel.

TECHNIQUE
Move your hips rearward, bend your forward knee slightly, and bend your upper body to the side and toward the handlebar. The elbow of your support arm should be straight, and you can push the bike a short distance to produce the maximum shoulder antepulsion. As your arm and your head come close together, the stretch will intensify and you will feel the muscle tension in your latissimus dorsi.

LEVEL	REPS	DURATION
BEGINNER	2	20 sec
INTERMEDIATE	2	25 sec
ADVANCED	2	25 sec

CAUTION

Start from a balanced position and be especially careful, as with all exercises that involve a broad movement in the shoulders. Reduce the intensity of the exercise if you feel any joint pain.

INDICATION

For all athletes, including cyclists who wish to maintain good overall flexibility, especially if they experience muscle tension in the latissimus dorsi, as commonly happens with swimmers.

UPPER LIMB STRETCHES

The arms, especially the wrists and elbows, are the body's main shock absorbers, because of their forward support on the bicycle, along with the cushioning provided by the palms of the hands, which, despite its essential nature, is less effective. In large measure, with their movement, these joints absorb the vibrations and minor shocks transmitted though the handlebar that are the product of riding on uneven surfaces. In addition, the extensor muscles of the elbow and the flexor carpi continually hold up a portion of the cyclist's weight, thereby keeping the body from falling forward. The work of supporting and absorbing shocks generates considerable fatigue and tension in parts of the arm and forearm muscles, such as the triceps and some of the flexor muscles of the wrist, so it is advisable to stretch them. Still, you must remember the muscles that spend a long time in a shortened position, even though they do not perform the same work load, such as the biceps brachii and some of the wrist extensor muscles, as they too need stretching in order to maintain their flexibility and optimal conditioning. The tension to which the arm muscles are subjected in the cycling sports varies notably as a function of the particular sport, from the up-and-down rocking on a road to absorbing the rough terrain in downhill race and dampening jumps and acrobatics in BMX. In any case, cyclists should remember the importance of the upper limbs in their sports.

BICEPS BRACHII

The long head of this muscle originates at the supraglenoid tuberculum, and its short head originates at the coracoid apophysis of the scapula. Both heads join and insert at the tuberosity of the radius. Their main function is bending the wrist, although they are also used in pronation of the elbow, and they play a role in antepulsion of the shoulder.

BRACHIALIS

This muscle originates at the anterior distal half of the diaphysis of the humerus, and it inserts at the tuberosity of the ulna, underneath the biceps brachii. Its function is bending the elbow.

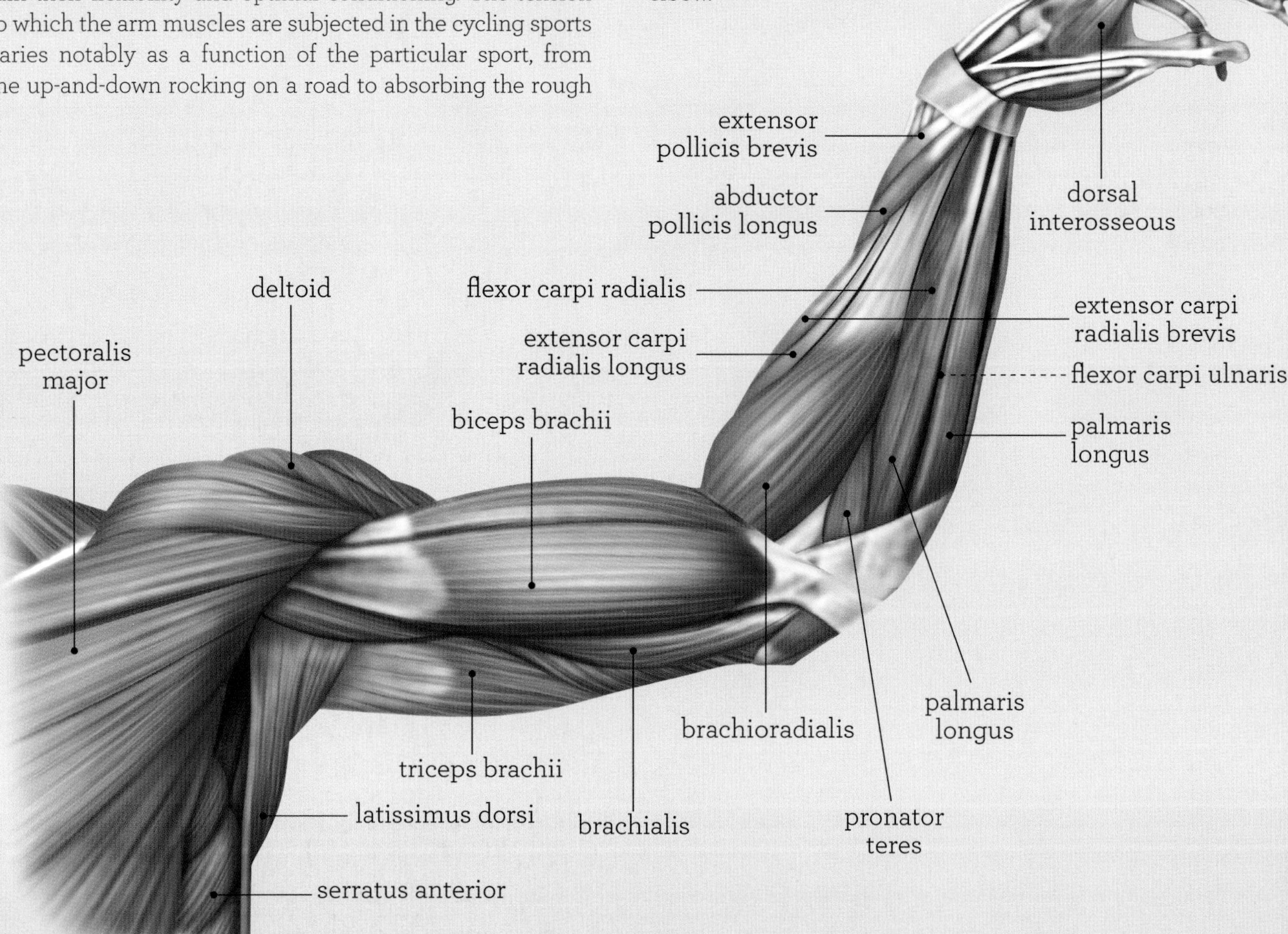

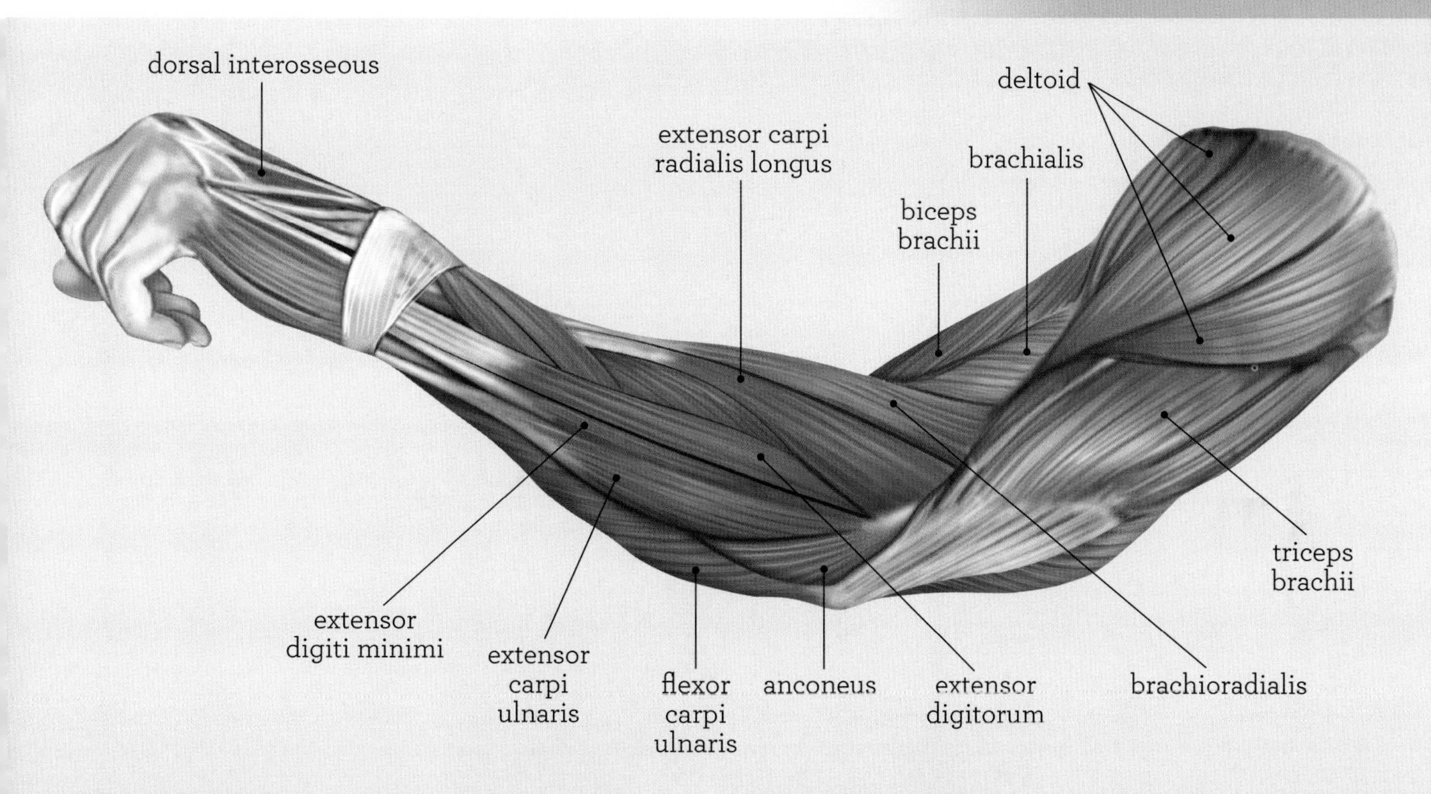

TRICEPS BRACHII
Its long head originates at the infraglenoid tuberculum of the scapula, its middle head in the distal two thirds of the diaphysis of the humerus on its rear face, and its lateral head at the proximal third of the diaphysis of the humerus, likewise on its rear face. All insert at the olecranon of the ulna. Their main function is straightening the elbow, with scant participation in retropulsion and the adduction of the shoulder.

EXTENSOR CARPI RADIALIS BREVIS
This muscle originates at the lateral epicondyle of the humerus, and it inserts at the base of the third metacarpal. Its main functions, which it shares with the extensor carpi radialis longus, are the extension and abduction of the wrist, although it also participates to a lesser degree in bending the elbow.

EXTENSOR CARPI RADIALIS LONGUS
This muscle originates at the lateral supracondyle crest of the humerus, and it inserts at the base of the second metacarpal.

EXTENSOR CARPI ULNARIS
This muscle originates at the lateral epicondyle of the humerus and the diaphysis of the ulna, and it inserts at the base of the fifth metacarpal. Its main functions are extension and abduction of the wrist.

FLEXOR CARPI ULNARIS
This muscle originates at the epitrochlea or medial epicondyle of the humerus, and at the olecranon of the ulna, and it inserts at the pisiform bones, the hamate, and the fifth metacarpal. Its functions are the flexion and adduction of the wrist.

FLEXOR CARPI RADIALIS
This muscle originates at the middle epicondyle of the humerus or epitrochlea, and it inserts at the base of the second metacarpal. Its functions are the flexion and abduction of the hand.

FLEXOR DIGITORUM SUPERFICIALIS
This muscle originates at the middle epicondyle of the humerus or epitrochlea, the coronoid apophysis of the ulna, and the anterior diaphysis of the radius, and it inserts at the second phalange of fingers two through five. Its functions are flexion of the wrist and the fingers, except for the thumb.

PALMARIS LONGUS
This muscle originates at the middle epicondyle of the humerus or epitrochlea, and it inserts at the palmar aponeurosis. Its main function is the flexion of the wrist, although it is used slightly in bending the elbow.

Wall Support with Twist

START
Stand next to a vertical support, such as a wall, a post, or a tree, but one step ahead of it. Your feet should be shoulder width apart. Rotate your upper body toward the vertical support without moving your feet and hold onto it with the closest hand, so that your thumb points upward and your palm faces front. Your grip should be at shoulder height, with a little bend in your elbow, and your arm and shoulder should be relaxed.

TECHNIQUE
Rotate your upper body away from the vertical support you are gripping, so that you produce shoulder retropulsion, and, above all, maximum extension in the elbow, the anterior face of which should remain facing downward. Once you reach this point, you should already feel the tension in the front of your arm, just above your elbow. If you do not feel it, you can increase the intensity of the stretch by slightly bending your knees and moving your center of gravity downward, without changing the grip in your original position.

Starting Position

LEVEL	REPS	DURATION
BEGINNER	2	25 sec
INTERMEDIATE	3	25 sec
ADVANCED	3	30 sec

CAUTION
It is important to conscientiously extend your elbow to the maximum, with its anterior face downward; otherwise, the stretch will be limited to the pectoral muscles rather than the biceps brachii.

INDICATION
For all cycling athletes, but mainly for track and road riders, especially if they ride intensively, because of the bend in the elbows during a major part of training and competition.

Pull from Rear

START
Stand with your back to the bicycle and hold the seat post with one hand, so that your thumb faces downward and your elbow is straight. You can use some other vertical support of a similar or greater height than the seat post. Place one leg ahead of the other, about one step apart. Both knees should be straight, with your back perpendicular to the ground. Your free hand can be placed onto the thigh of the corresponding leg for added comfort while doing the exercise.

TECHNIQUE
Bend both knees so that your center of gravity moves downward considerably. Your upper body should remain perpendicular to the ground, and your grip should remain in its original position. As your center of gravity lowers, you will produce shoulder retropulsion. Keep your elbow straight with its anterior face downward. This will produce the sensation of stretching in the front part of your arm, just above the elbow.

LEVEL	REPS	DURATION
BEGINNER	2	25 sec
INTERMEDIATE	3	25 sec
ADVANCED	3	30 sec

CAUTION
If you do this stretch with the bicycle, make sure it is stable before you start the exercise. If you do not feel the stretch, repeat the exercise with a higher anchor point.

INDICATION
For all cyclists, especially track and road riders who use aero bars for long periods of time during training or competition, because of the bend in their elbows.

Rear Elbow Pull

START

Stand with your upper body perpendicular to the ground. Raise one arm so that it is in line with your torso, and hold the elbow of this arm with the opposite hand so that your forearm is on top of your head. Your head should be tilted slightly forward to provide a better grip and avoid interfering with the performance of the exercise. At this point, the elbow being held can be partially bent, with your forearm relaxed.

TECHNIQUE

Bend the elbow of the first arm as far as possible, as if you were trying to reach the middle of your back with your hand, while the opposite hand pulls the elbow rearward, trying to move it behind your head. As the elbow of the first arm bends and increases the pull of the opposite hand, you will feel the tension in the back of your arm, especially in the area closest to your shoulder. This is a sign that you are doing the stretch properly.

Starting Position

LEVEL	REPS	DURATION
BEGINNER	2	20 sec
INTERMEDIATE	2	25 sec
ADVANCED	3	25 sec

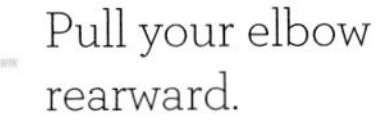

triceps brachii

teres major

latissimus dorsi

CAUTION

Even though this exercise entails no major risk, you need to be careful with the intensity and the speed with which you pull on your elbow, because your shoulder is in maximum antepulsion and you could experience some discomfort.

INDICATION

For all cyclists without exception, because of the constant work that the triceps brachii performs as support on the handlebar. This effort is prolonged and sustained in road riding, and more irregular in mountain biking.

Rearward Pull with Towel

Bend the elbow and the shoulder of the raised arm as far as possible.

triceps brachii

teres major

latissimus dorsi

Starting Position

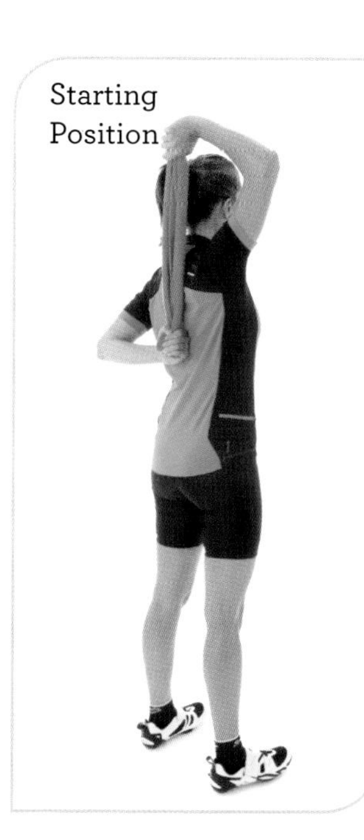

START

For this exercise, you will need a towel for pulling on both ends. Stand and hold the towel with one hand over your head so that your fist is uppermost. Bend your elbow slightly so that the towel hangs down behind your back. With your other hand, grab the loose end of the towel hanging down behind you, around 16 inches (40 cm) from the first hand, so that the elbow of the second hand is bent.

TECHNIQUE

Straighten the elbow of the second arm while maintaining both grips, which will necessitate maximum flexion in the upper elbow and similarly maximum antepulsion in the shoulder on the same side. Throughout the exercise, you will have to keep the towel or other item stretched tight in order to produce the stretch properly. You will feel tension in the back of the raised arm, and more intensely in the area closest to the shoulder joint.

LEVEL	REPS	DURATION
BEGINNER	2	20 sec
INTERMEDIATE	2	25 sec
ADVANCED	2	30 sec

CAUTION

If you use an item longer than 16 inches (40 cm), try to keep your hands no farther apart than this distance, and pay attention to any discomfort in your shoulder. If you experience any discomfort, you should reduce the intensity of the stretch.

INDICATION

For all cycling athletes, especially those who do lengthy workouts or competitions, ride over rough terrain, or perform jumps, because of the special involvement of the triceps brachii in supporting the body on the handlebars and in absorbing shocks and vibrations.

Pull with Wrist Extension

START
Hold one hand at waist height and slightly in front of your body. Your palm will face forward, with your fingers pointing downward. With your free hand, hold the first one by placing your four fingers on the palm and your thumb on the back of the hand. The forearm of the pulling hand will cross in front of your body, and your other forearm will rest against your side.

TECHNIQUE
Straighten the elbow of the side being stretched so that your hand moves forward. The holding hand will maintain and accentuate the pull on the other one, so that the wrist of the latter will extend as far as possible and produce tension in the flexor muscles, including several of the epitrochlear muscles. When you reach a sufficient degree of stretch, you will feel the tension in the middle proximal area of your forearm.

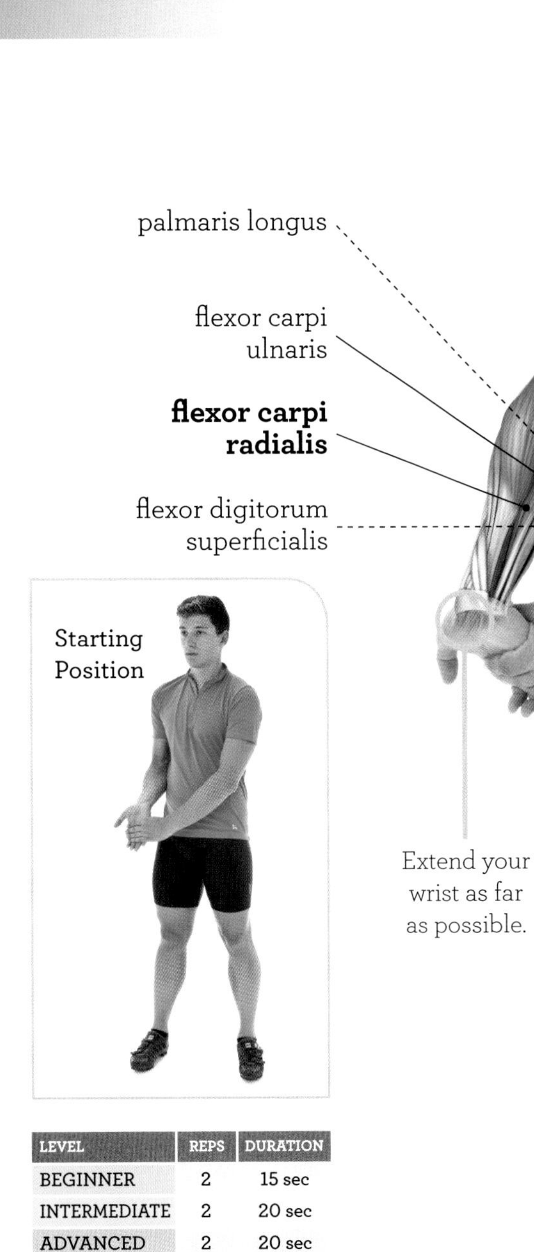

LEVEL	REPS	DURATION
BEGINNER	2	15 sec
INTERMEDIATE	2	20 sec
ADVANCED	2	20 sec

CAUTION

Pull carefully on your wrist, because the part that is involved in the stretch is very sensitive to sudden tension.

INDICATION

For all cyclists, especially road racers, because of the constant tension to which the wrist muscles are subjected, mainly in the basic grip on top of the bars, or with hands placed just before the curve of the bars.

Inverted Bilateral Support on Handlebars

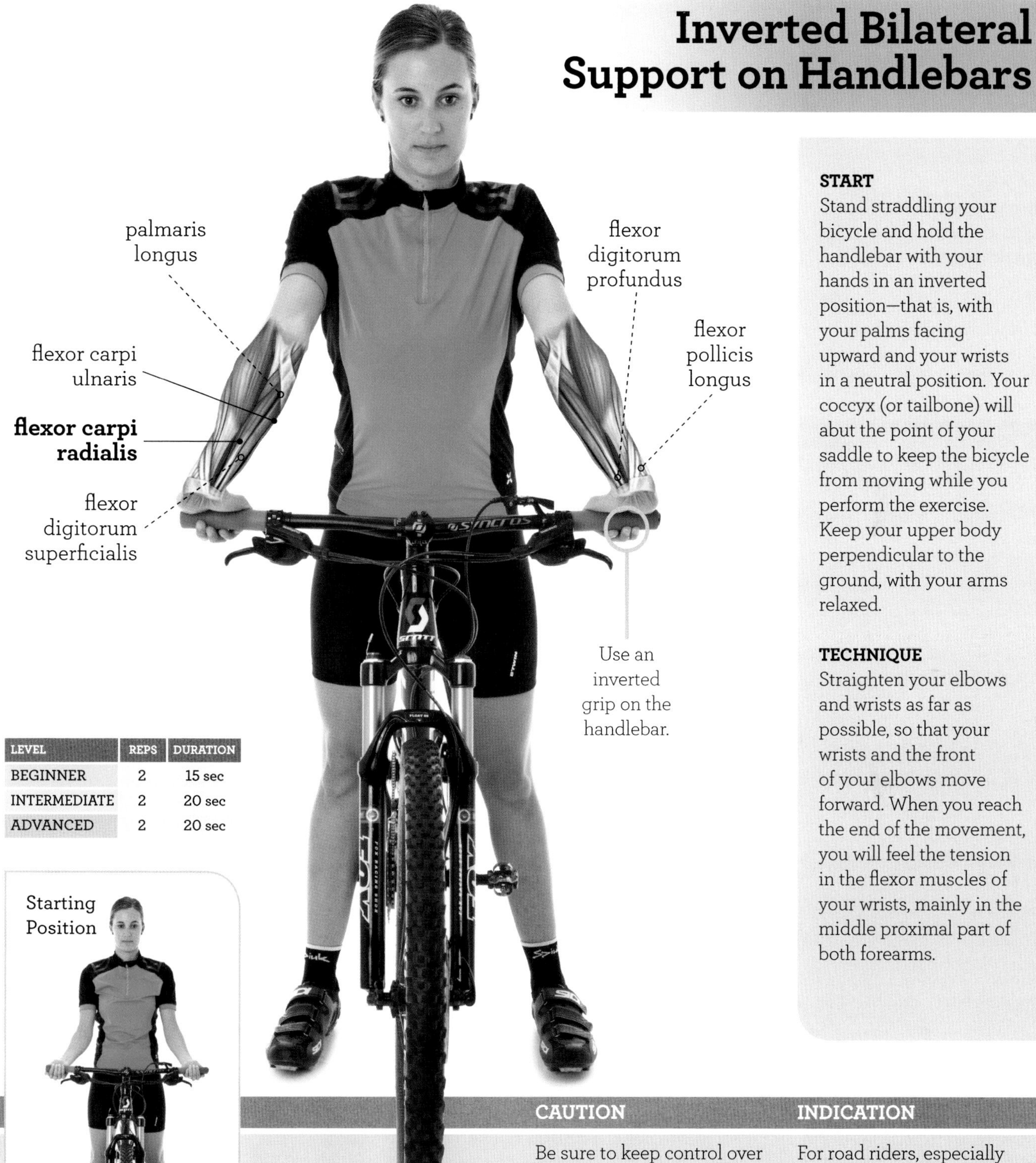

START
Stand straddling your bicycle and hold the handlebar with your hands in an inverted position—that is, with your palms facing upward and your wrists in a neutral position. Your coccyx (or tailbone) will abut the point of your saddle to keep the bicycle from moving while you perform the exercise. Keep your upper body perpendicular to the ground, with your arms relaxed.

TECHNIQUE
Straighten your elbows and wrists as far as possible, so that your wrists and the front of your elbows move forward. When you reach the end of the movement, you will feel the tension in the flexor muscles of your wrists, mainly in the middle proximal part of both forearms.

LEVEL	REPS	DURATION
BEGINNER	2	15 sec
INTERMEDIATE	2	20 sec
ADVANCED	2	20 sec

Starting Position

CAUTION

Be sure to keep control over the bike, and if you feel any discomfort in your wrists, reduce the intensity of the stretch.

INDICATION

For road riders, especially after long workouts gripping the tops of the bars or just before the drops, because these grips place the wrist extensors under tension.

Pull with Wrist Bend

START
Stand up straight. Place both hands in front of you by bending your shoulders and hold them in a position of pronation—in other words, palms down. Hold one hand with the other so that your four fingers are on the back, with your thumb in contact with the palm, like a clamp. Your elbows should be straight, and the wrist of the side being stretched is in a neutral position.

TECHNIQUE
Pull the hand you are holding downward and toward the outside. This will produce a flexion and an outer rotation of the wrist, while you keep the arm of the side being stretched in its original position. As the movement progresses, you will feel the stretch in the lateral proximal part of your forearm, which you will see in the final position.

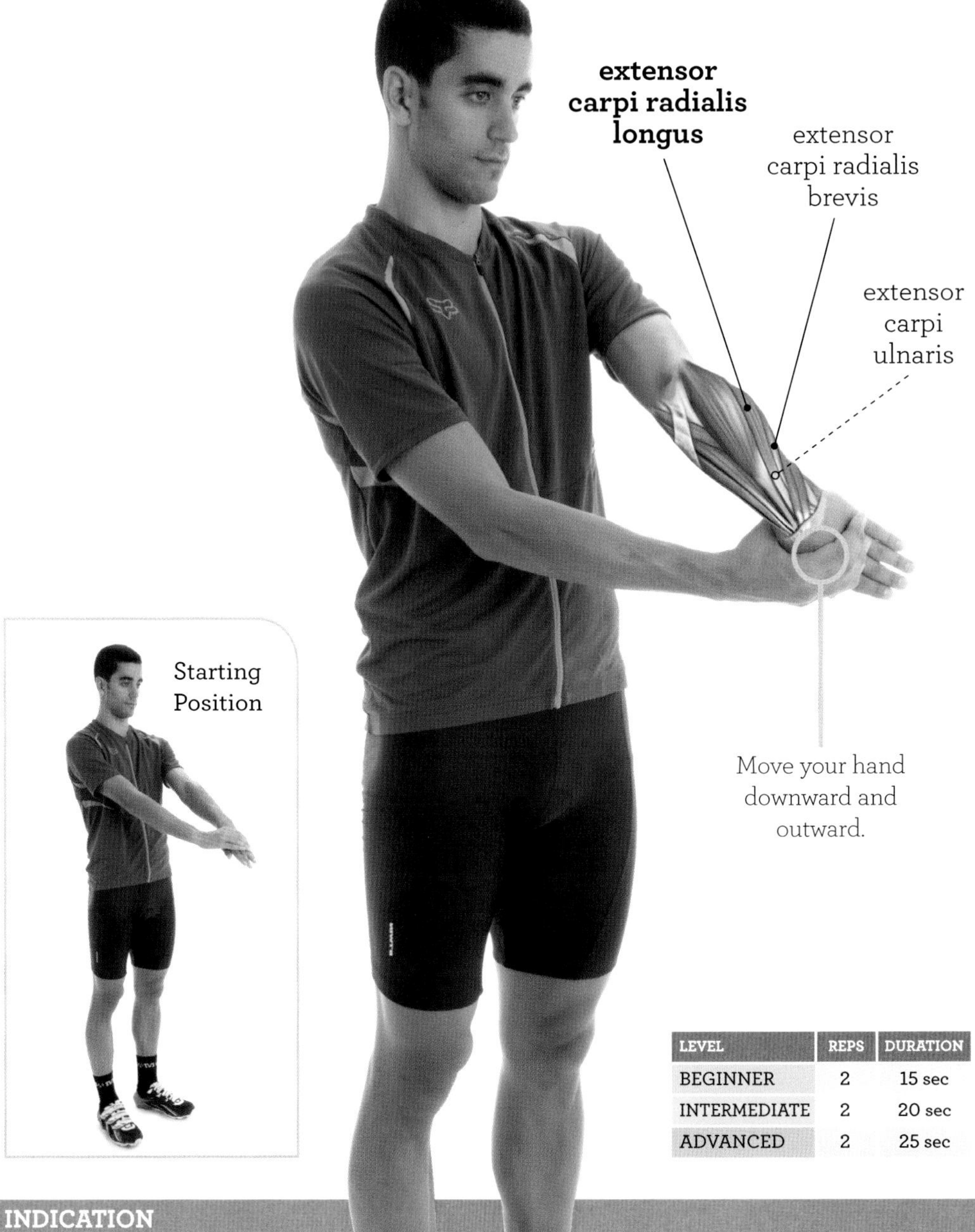

LEVEL	REPS	DURATION
BEGINNER	2	15 sec
INTERMEDIATE	2	20 sec
ADVANCED	2	25 sec

CAUTION

Perform the pull in a gradual and controlled manner, and avoid forcing the wrist. Remember that the stretch should produce a little discomfort, but not pain.

INDICATION

For cyclists who hold the bars on the top or just before the drops for prolonged periods, because of the imbalances that may result between the wrist flexor and extensor muscles.

Bilateral Back Support on Handlebars

LEVEL	REPS	DURATION
BEGINNER	2	15 sec
INTERMEDIATE	2	20 sec
ADVANCED	2	25 sec

START
Straddle your bicycle. Move your hands to the handlebar in a position of pronation and rest the upper part of your knuckles below the bars. Your elbows should be partly bent, with your wrists in a neutral position. The point of your saddle should touch your coccyx (tailbone) so that the bicycle remains stationary while you do the exercise.

TECHNIQUE
Straighten your elbows and lift your arms slightly while keeping the upper part of your knuckles in contact with the handlebar. This will produce maximum flexion in your wrists, which should be rotated slightly so that your fingers point downward and outward. This will maximize the effect of the stretch on the wrist extensor muscles and several epicondyle muscles.

CAUTION
Don't use too much force in this stretch, because of the relative fragility of the wrists, and, in any case, avoid stretching to the point of feeling pain.

INDICATION
For all cyclists, given that the most commonly used grip is the basic one on the transverse part of the handlebar. This commonly involves a certain extension of the wrist, and thus a sustained, shortened position of the muscles that perform this function.

WRIST AND HAND STRETCHES

The hand is the contact point between the cyclist's upper limbs and the handlebar of the bicycle, and the wrist is the first joint to absorb vibrations and shocks. In addition, the fingers continually work to achieve a good, firm grip, and they are responsible for working both the gear shifters and the brake levers during races and training. This makes these parts of the body very important in cycling, even though they are commonly neglected in training. In the long run, pushing hard for a long time on the handlebar and the maximum extension of the wrist in certain grips can cause compression of the median or the ulnar nerve, which runs along the front part of the wrist. Compression of the medial nerve can cause pain, tingling, or numbness in the fingers, and compression of the ulnar nerve can produce these in the ring and little fingers. As a result, it is a good idea to frequently change grip, move your wrists, and even do stretches regularly. The flexor muscles of the fingers also are subjected to great tension, due to the ongoing necessity of not only holding the handlebar, but also gripping it firmly and working the gear shifters and the brakes, so it is a good idea to stretch them to avoid cramps, pain, and even reduced mobility, especially if the bike is ridden regularly and for a long time in any sport.

ADDUCTOR POLLICIS
The transverse head of this muscle originates at the third metacarpal, and its oblique head, at the second and third metacarpals and the capitate bone. Both heads join and insert at the base of the first proximal phalange. Their main functions are adduction in the carpometacarpal joint and flexion in the metacarpophalangeal joint of the thumb.

EXTENSOR POLLICIS BREVIS
This muscle originates at the posterior face of the radius and the interosseous membrane, and it inserts at the base of the first proximal phalange. Its functions are abduction of the wrist and extension of the thumb in its carpometacarpal and metacarpophalangeal joints.

EXTENSOR DIGITORUM
This muscle originates at the lateral epicondyle of the humerus, and it inserts at the phalanges of the four fingers through the dorsal aponeurotic expansion. Its main functions are the extension of the wrist and the four fingers in the metacarpophalangeal and interphalangeal joints.

EXTENSOR INDICIS
This muscle originates at the rear face of the ulna and the interosseous membrane, and it inserts at the middle and distal phalanges of the index finger through the dorsal aponeurotic expansion of the index finger. Its functions are the extension of the wrist and the index finger at the metacarpophalangeal and interphalangeal joints.

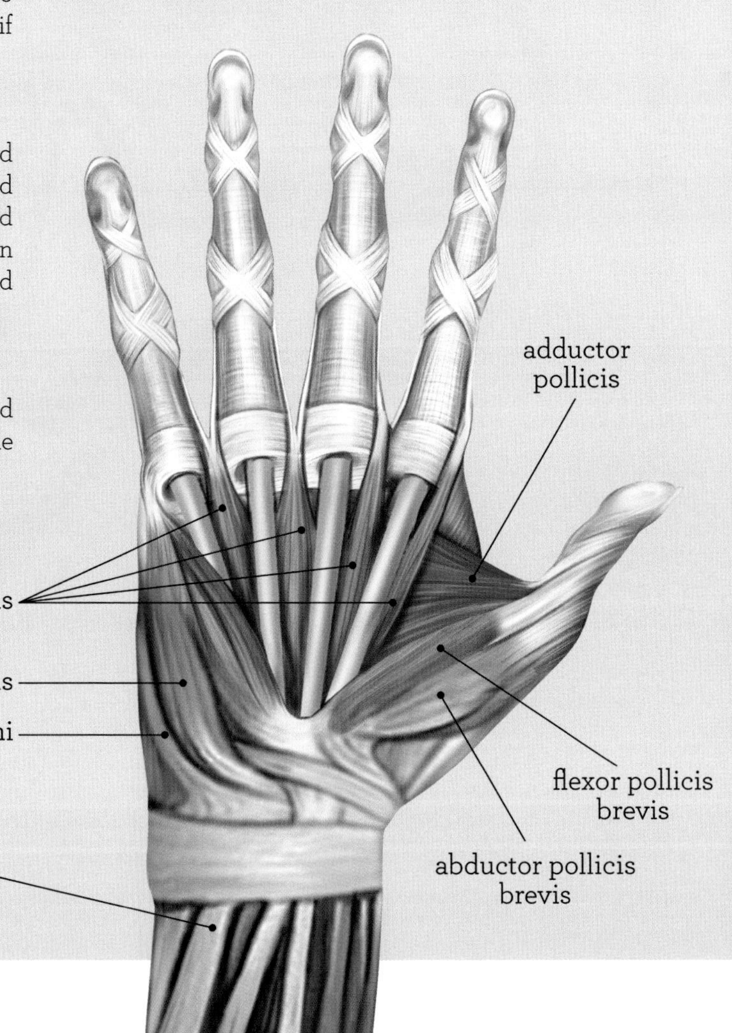

EXTENSOR DIGITI MINIMI
This muscle originates at the lateral epicondyle of the humerus, and it inserts at the phalanges of the little finger through the expansion of its dorsal aponeurosis. Its main function is the extension of the little finger at its metacarpophalangeal and interphalangeal joints, as well as the abduction of this finger and the wrist.

EXTENSOR POLLICIS LONGUS
This muscle originates at the rear face of the ulna and the interosseous membrane, and it inserts at the first distal phalange. Its functions are extension and abduction of the wrist, plus extension of the thumb at its metacarpophalangeal and interphalangeal joints.

FLEXOR POLLICIS BREVIS
This muscle originates at the flexor retinaculum and the capitate and the trapezium bones, and it inserts at the base of the first proximal phalange. Its function is flexion in the metacarpophalangeal joint of the thumb.

FLEXOR POLLICIS LONGUS
This muscle originates at the front face of the radius and the interosseous membrane, and it inserts at the first distal phalange. Its functions are flexion of the thumb in its carpometacarpal, metacarpophalangeal, and interphalangeal joints, as well as flexion and abduction of the wrist.

FLEXOR DIGITORUM PROFUNDUS
This muscle originates at the front face of the ulna and the interosseous membrane, and it inserts at the palmar face of phalanges two through five. Its main function is flexion of the wrist and the metacarpophalangeal and interphalangeal joints of fingers two through five.

FLEXOR DIGITORUM SUPERFICIALIS
This muscle originates at the middle epicondyle of the humerus, the coronoid apophysis of the ulna, and the proximal anterior surface of the radius, and it inserts at the middle phalanges of fingers two through five. Its function is flexion of the wrist and the metacarpophalangeal and proximal interphalangeal joints of fingers two through five.

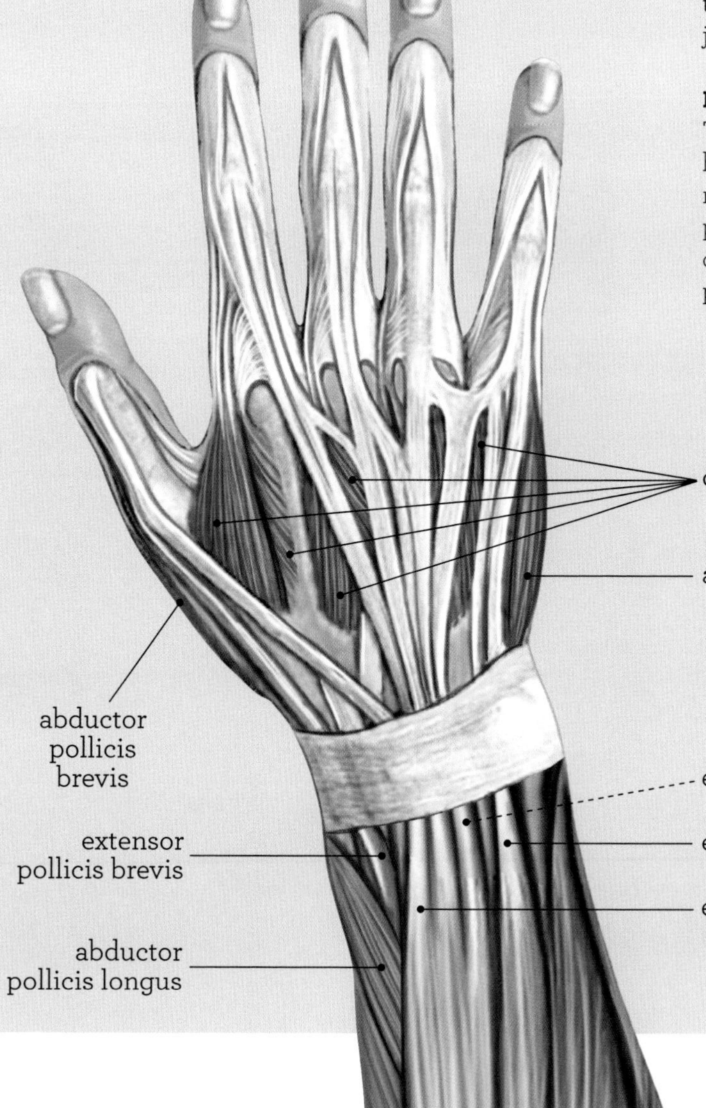

Finger Extension

START
Stand and place your hands in supination right in front of your upper body, but without touching it, while keeping your elbows bent at 90° and your shoulders relaxed. Fold the fingers of both hands together while keeping them in supination, so that your thumbs are the only digits that are not crossed.

TECHNIQUE
Turn your palms downward without disturbing the original position of your fingers or separating them. Straighten your elbows and flex your shoulders slightly so that your wrists and fingers are extended. Your palms should face forward and downward. The tension will gradually become apparent in the palms and the anterior face of your fingers.

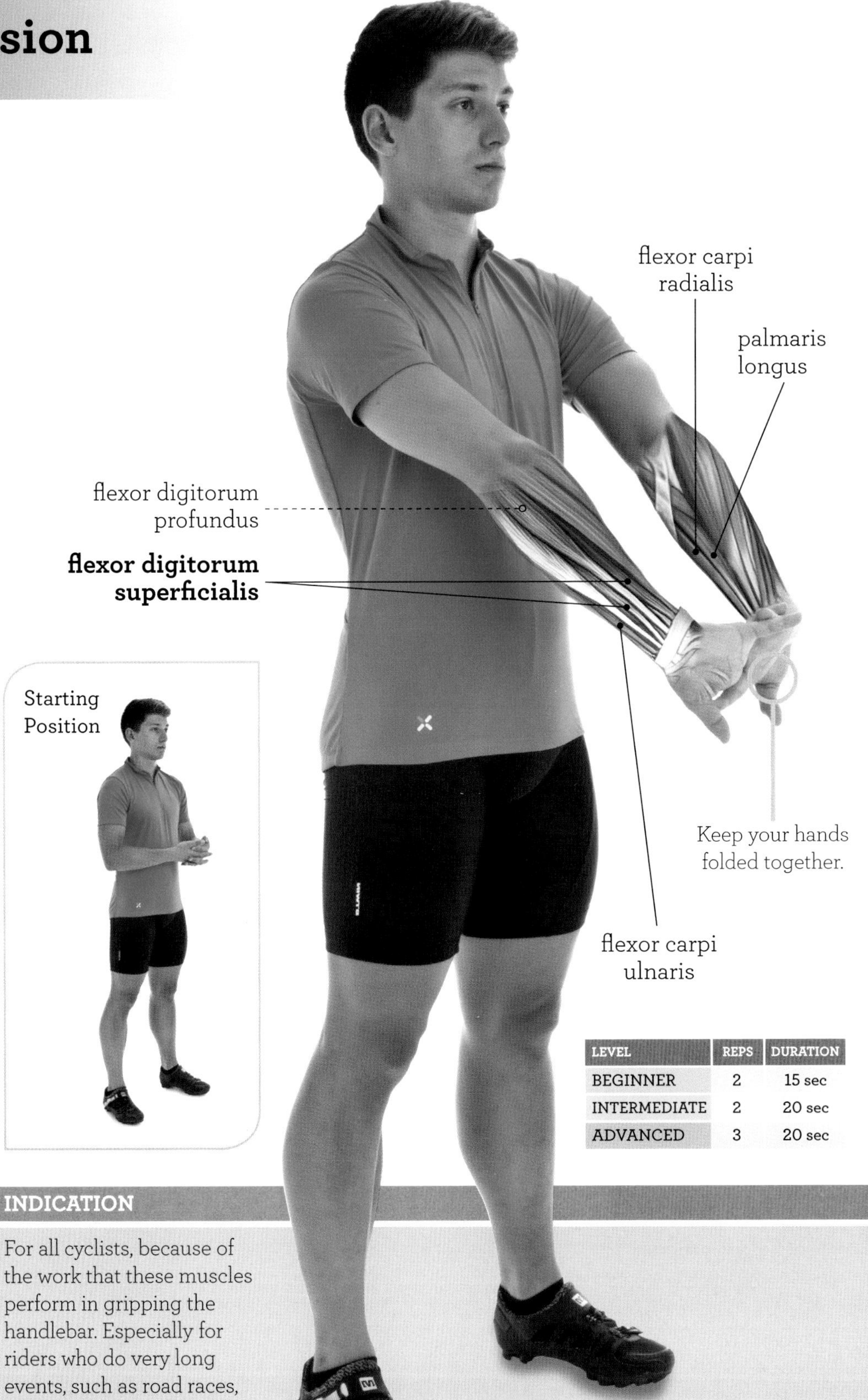

LEVEL	REPS	DURATION
BEGINNER	2	15 sec
INTERMEDIATE	2	20 sec
ADVANCED	3	20 sec

CAUTION
Be careful when applying pressure to your fingers, given the relative fragility of their joints, and do not exceed the recommended times and ranges.

INDICATION
For all cyclists, because of the work that these muscles perform in gripping the handlebar. Especially for riders who do very long events, such as road races, and who use the brakes more intensively, such as mountain bikers.

Prayer Position

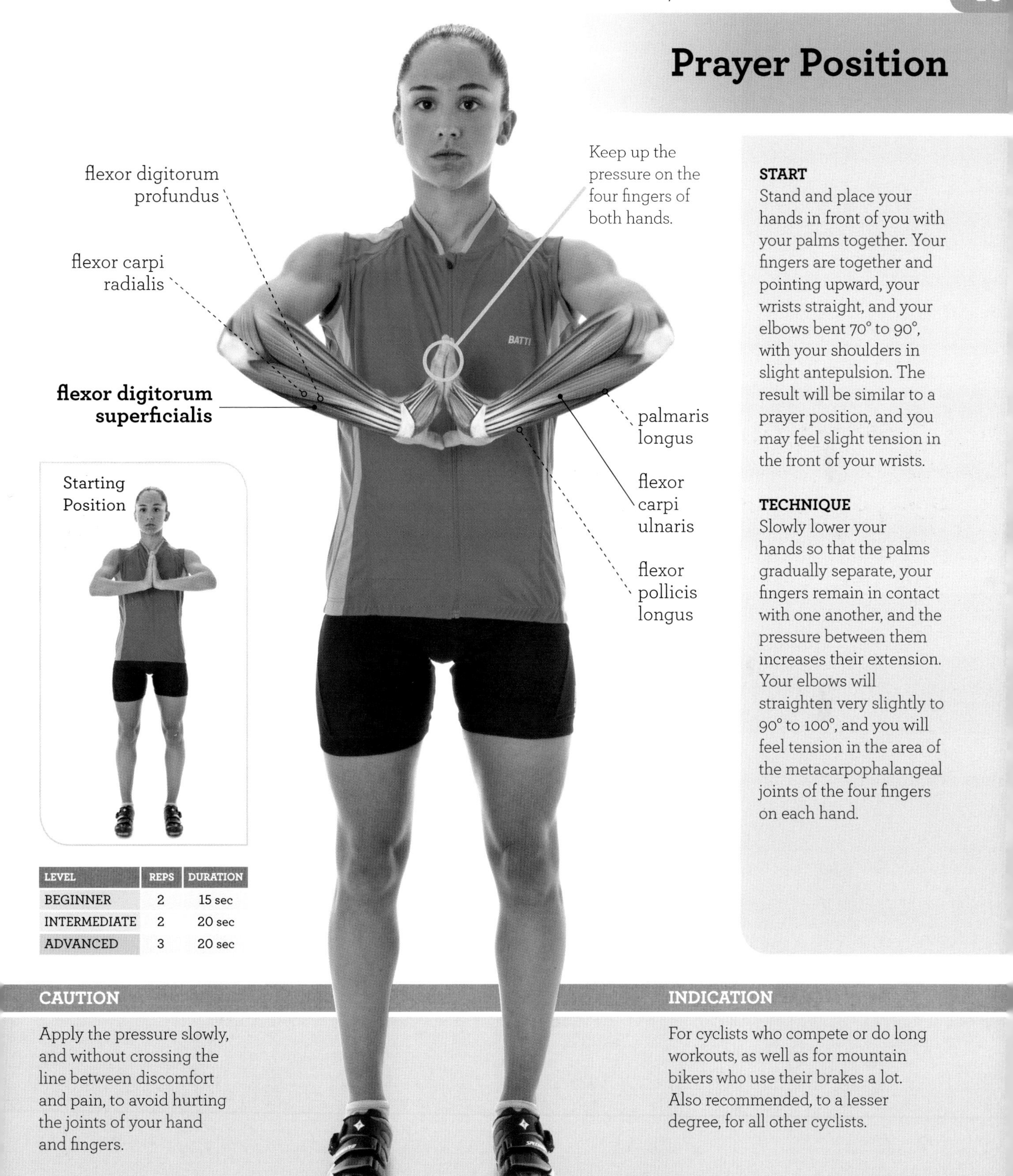

LEVEL	REPS	DURATION
BEGINNER	2	15 sec
INTERMEDIATE	2	20 sec
ADVANCED	3	20 sec

START
Stand and place your hands in front of you with your palms together. Your fingers are together and pointing upward, your wrists straight, and your elbows bent 70° to 90°, with your shoulders in slight antepulsion. The result will be similar to a prayer position, and you may feel slight tension in the front of your wrists.

TECHNIQUE
Slowly lower your hands so that the palms gradually separate, your fingers remain in contact with one another, and the pressure between them increases their extension. Your elbows will straighten very slightly to 90° to 100°, and you will feel tension in the area of the metacarpophalangeal joints of the four fingers on each hand.

CAUTION

Apply the pressure slowly, and without crossing the line between discomfort and pain, to avoid hurting the joints of your hand and fingers.

INDICATION

For cyclists who compete or do long workouts, as well as for mountain bikers who use their brakes a lot. Also recommended, to a lesser degree, for all other cyclists.

Thumb Pull

START
Stand with your elbows bent to about 90° and your hands right in front of your stomach. Place one hand with the palm facing you, the four fingers curled, and your thumb upward, similar to the thumbs-up sign. Grab the thumb with your other hand, surrounding it with your fingers in a fist.

TECHNIQUE
Use the hand holding your thumb to push it rearward and reach the maximum extension in the carpomatacarpal, metacarpophalangeal, and interphalangeal joints of the thumb. As you increase the pull on the thumb, you will feel the tension of the stretch in the thenar eminence and the area between the thumb and index finger.

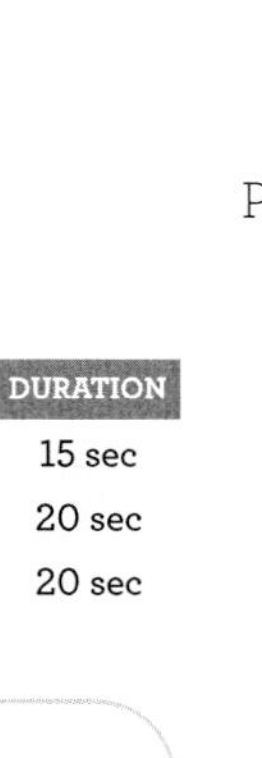

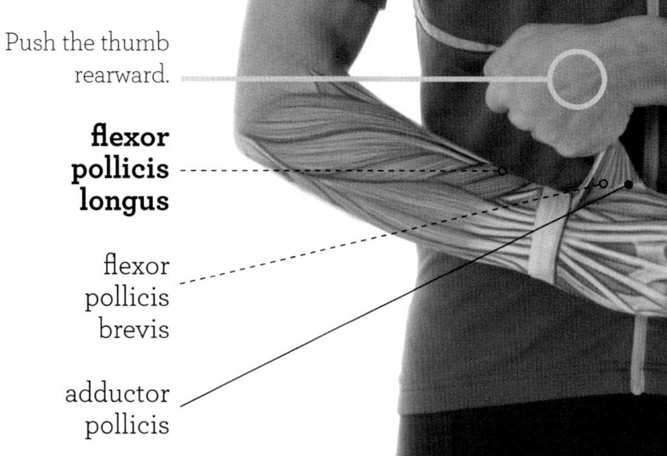

LEVEL	REPS	DURATION
BEGINNER	2	15 sec
INTERMEDIATE	2	20 sec
ADVANCED	2	20 sec

Starting Position

CAUTION
As with all stretches involving the fingers, you should apply moderate, gradual tension to avoid harming the muscles of your hand.

INDICATION
For cyclists whose sports involve prolonged, frequent use of the bicycle, significant pressure in gripping the handlebars, or intensive use of the brakes, as in downhill events, time racing, and cross-country events.

Rhombus Position

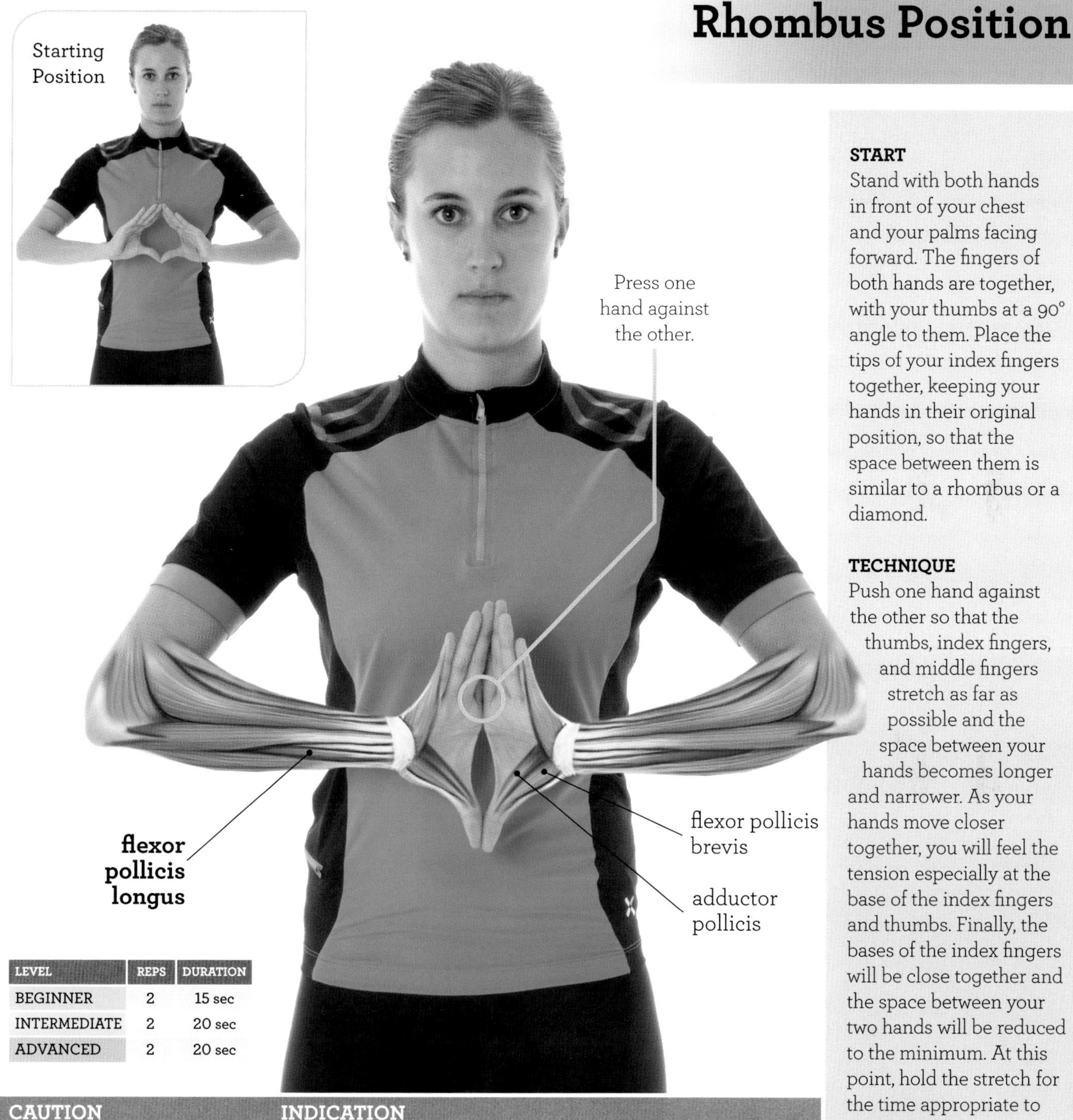

START
Stand with both hands in front of your chest and your palms facing forward. The fingers of both hands are together, with your thumbs at a 90° angle to them. Place the tips of your index fingers together, keeping your hands in their original position, so that the space between them is similar to a rhombus or a diamond.

TECHNIQUE
Push one hand against the other so that the thumbs, index fingers, and middle fingers stretch as far as possible and the space between your hands becomes longer and narrower. As your hands move closer together, you will feel the tension especially at the base of the index fingers and thumbs. Finally, the bases of the index fingers will be close together and the space between your two hands will be reduced to the minimum. At this point, hold the stretch for the time appropriate to your level.

LEVEL	REPS	DURATION
BEGINNER	2	15 sec
INTERMEDIATE	2	20 sec
ADVANCED	2	20 sec

CAUTION

If you feel any pain, reduce the intensity of the exercise or stop. Again, remember the relative fragility of the fingers.

INDICATION

For all cyclists, but especially for those who spend lots of time on the bicycle or need to grip the handlebar firmly because they ride over rough terrain.

Thumb Support

START
Straddle your bicycle with the handlebars straight. Rest the tips of your thumbs on the top of the handlebars, with your fists closed, your elbows straight, and a slight flexion in your shoulders. Exert no pressure in this first phase and keep your upper limbs relaxed.

TECHNIQUE
Lower your hands while keeping your thumbs on top of the handlebars, so that the thumbs support the tension generated by the weight of your arms. You probably will feel tension at the base of your thumbs and the thenar eminence. If not, you can apply additional pressure on the thumb supports.

flexor pollicis longus

adductor pollicis

flexor pollicis brevis

LEVEL	REPS	DURATION
BEGINNER	2	15 sec
INTERMEDIATE	2	20 sec
ADVANCED	2	20 sec

CAUTION

Never cross the threshold of pain, and remember to apply pressure increases very gradually, because of the size of the muscles affected by this stretch.

INDICATION

For cyclists who experience tension or fatigue in the flexor muscles of the fingers due to long sessions on the bicycle or great pressure in the fingers on the handlebars, in particular in riding the roads and in various mountain biking events.

Wrist and Finger Bend

Starting Position

extensor digitorum

extensor pollicis

extensor digiti minimi

extensor pollicis longus

Bend the wrist and the hand that are bring stretched.

extensor indicis

LEVEL	REPS	DURATION
BEGINNER	2	15 sec
INTERMEDIATE	2	20 sec
ADVANCED	2	20 sec

START

Bend your elbows and hold your hands in front of your upper body. The fingers of one hand, except for the thumb, will be bent like claws, and you will hold the other hand over fingers two through five and cover them. This way, the holding hand will cover the fingers of the other one, the wrist of which will be extended, neutral, or at some intermediate point.

TECHNIQUE

Using the holding hand, push the other hand downward while keeping both elbows bent. This way, the bend in the wrist will increase and the fingers will remain bent to the maximum, both in their metacarpophalangeal and interphalangeal joints, except for the thumb. As you continue performing the technique, you will feel the tension in the back of the hand that is being stretched.

CAUTION

Don't try to achieve a high degree of flexibility quickly, but rather try to progress gradually in the stretch and in successive sessions, so you don't exceed the safety limits that every exercise should observe.

INDICATION

For all cyclists, but especially for road riders who keep their hands on top of the handlebars or just before the drops, use the gear shifters and brakes a lot, often ride with their index and middle fingers on the brake levers, or spend long times on the bicycle in training or competition.

grip
SHIMANO

HIP AND LOWER LIMB STRETCHES

HIP STRETCHES

The muscles that move the hips play an essential role in cycling, because, along with the muscles that move the knee and the ankle, they are the ones that are used in pedaling, and thus they are the main ones involved in moving the bike forward. The muscles that extend the hip are the ones that do the most work. Hip extension requires adequate preparation, and cyclists often experience muscle strain, such as chronic strain of the gluteal muscles, which especially affects climbers and which can be treated with an appropriate program of stretching, among other therapeutic measures. In dealing with the hip flexors, although they work at a lower level of intensity, it is appropriate to keep in mind their relatively shortened position in most cycling, because the hips are not extended completely in pedaling, so these muscles need to be stretched regularly. Other common problems in cycling that manifest themselves in the hip muscles may arise from the repetition of pedaling, improper adjustment of the bicycle, and a lack of flexibility in certain muscles. In this area, trochanteric bursitis or iliotibial tract syndrome are common problems in cycling that affect a great many devotees. Treatment includes stretching the tensor fasciae latae, one of the muscles that contribute to the abduction movement of the hip. Other cycling-related problems are found in the hip and genital areas, such as irritation of the sciatic tuberosity and numbness in the genitals, but the origin of these is not muscular, and their alleviation is connected more to the adoption of a proper position on the bicycle, the choice of a certain saddle, and its correct adjustment.

ADDUCTOR MAGNUS

This muscle originates at the lower branch of the pubis, the sciatic branch, and the sciatic tuberosity, and it inserts at the linea aspera of the femur at the proximal two thirds of the diaphysis. Its main function is adduction of the hip joint, although it also participates to a lesser degree in its extension, flexion, and rotation.

ADDUCTOR LONGUS

This muscle originates at the upper branch of the pubis and inserts at the linea aspera of the femur at the middle third of the diaphysis. Its main function is adduction of the hip, although it participates to a lesser degree in its flexion and rotation.

ADDUCTOR BREVIS

This muscle originates at the lower branch of the pubis, and it inserts at the linea aspera of the femur at the proximal third of the diaphysis. Its function is adduction of the hip, and, to a lesser degree, its flexion and rotation.

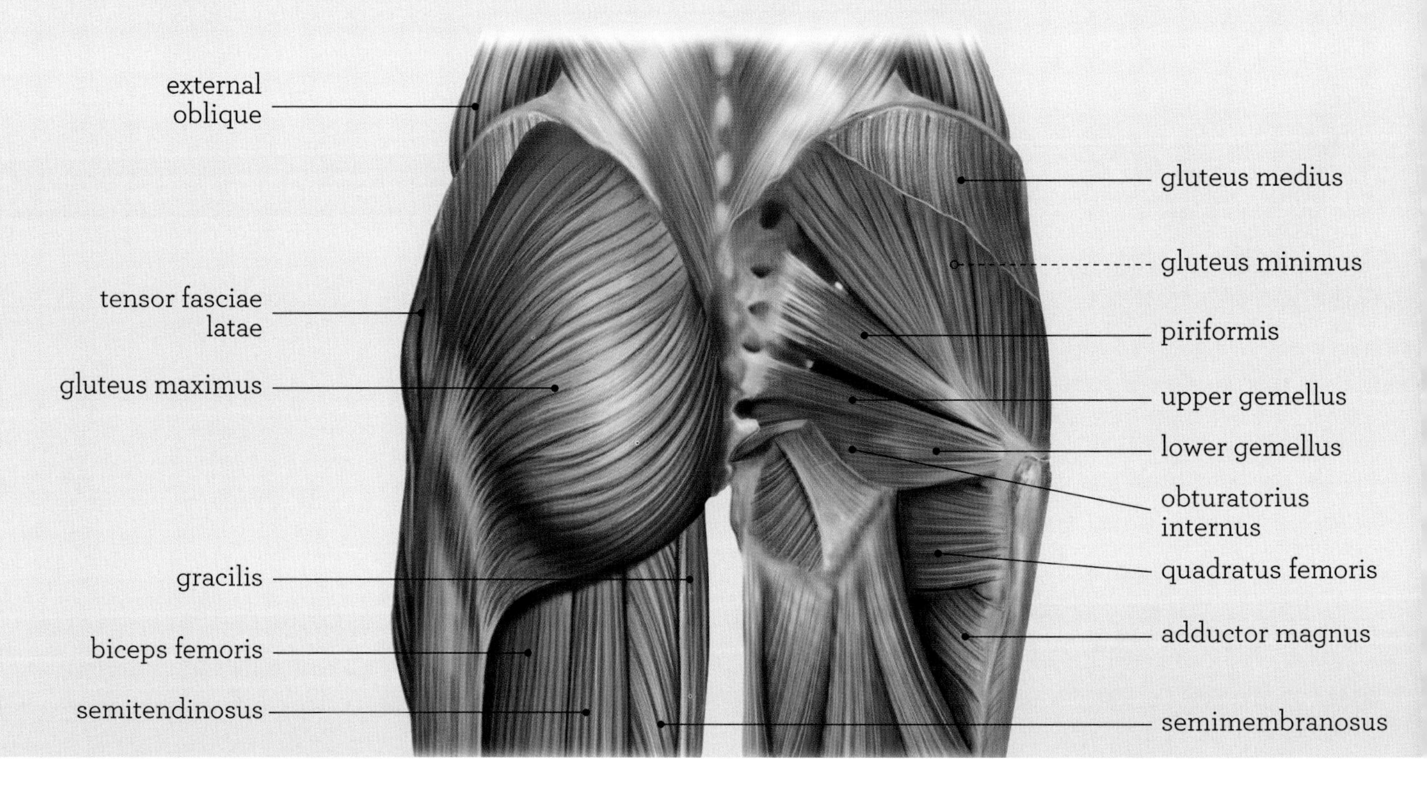

GLUTEUS MAXIMUS
This muscle originates at the posterior face of the sacral bone, the posterior face of the ilium, the thoracolumbar fascia, and the sacrotuberous ligament, and it inserts at the iliotibial tract and the gluteal tuberosity of the femur. Its main functions are extension and abduction of the hip, although it also participates in its external rotation and adduction.

GLUTEUS MEDIUS
This muscle originates at the posterior face of the ilium, and it inserts at the trochanter major of the femur. Its main function is abduction of the hip, although it participates in its flexion, extension, and rotation, both external and internal.

GLUTEUS MINIMUS
This muscle originates at the posterior face of the ilium, below the origins of the gluteus maximus and gluteus medius. Its insertion is at the trochanter major of the femur, and it shares functions with the gluteus medius.

ILIACUS
This muscle originates at the iliac fossa, and it inserts at the trochanter minor of the femor. Its main functions are flexion and external rotation of the hip.

PECTINEUS
This muscle originates at the upper branch of the pubis, and it inserts at the linea aspera and the linea pectinea femoris. Its main function is adduction of the hip, with lesser participation in its flexion and external rotation.

PIRIFORMIS
This muscle originates at the anterior face of the sacral bone, and it inserts at the trochanter magnus of the femur. Its main functions are abduction, extension, and external rotation of the hip.

PSOAS MAJOR
This muscle originates at the intervertebral bodies and discs of vertebrae T12 through L4, as well as at the transverse apophyses of vertebrae L1 through L5. It inserts at the trochanter minor of the femur, and it shares functions with the iliacus muscle.

GRACILIS
This muscle originates at the lower branch of the pubis, and it inserts at the proximal middle part of the tibia. Its main functions are adduction and flexion of the hip, plus flexion of the knee.

TENSOR FASCIAE LATAE
This muscle originates at the upper anterior iliac spine, and it inserts at the iliotibial tract. Its main functions are abduction, flexion, and internal rotation of the hip.

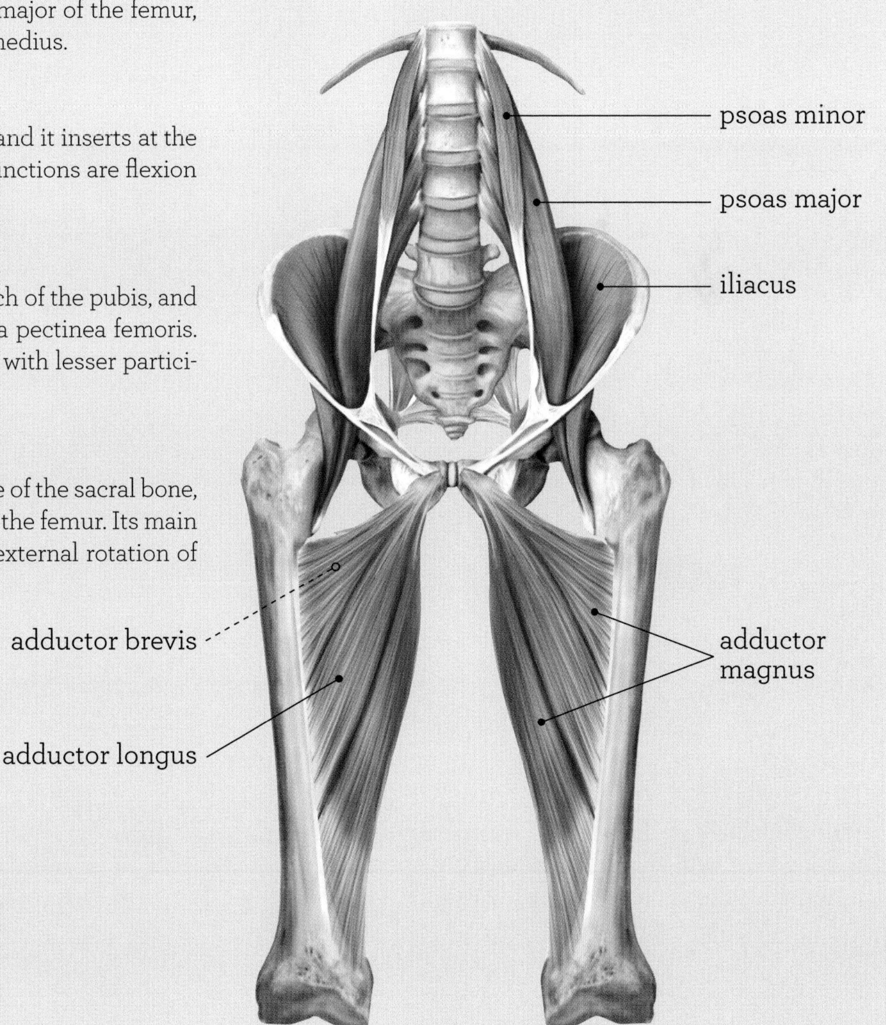

V-seat

START
Sit on the ground or a mat with your legs in front of you and your knees straight. Separate your legs slightly, move your arms forward, and place your palms on the insides of your calves. You will have to lean your upper body slightly forward.

TECHNIQUE
Separate your legs as much as possible and push with your hands to reach the widest possible angle, largely toward the end of the range of motion, at which point the action of the abductors will reduce the effectiveness. A feeling of tension will become evident in the inner area of both thighs and the groin, which proves the effectiveness of the stretch.

Place your palms at the middle and inside of your calves.

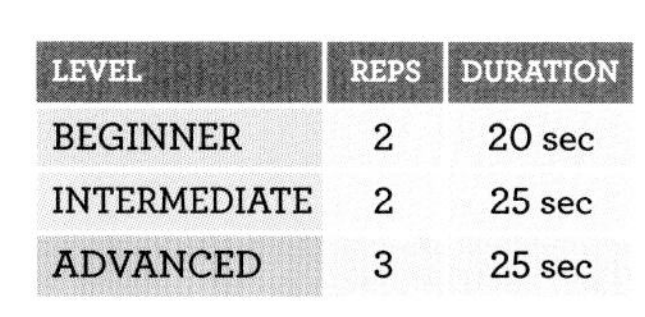

LEVEL	REPS	DURATION
BEGINNER	2	20 sec
INTERMEDIATE	2	25 sec
ADVANCED	3	25 sec

Starting Position

CAUTION

Do the movement slowly and try to avoid pain. At the same time, avoid bending your lumbar vertebrae excessively and the resulting discomfort in this area.

INDICATION

For all cyclists, regardless of their sports, in order to contribute to the best possible overall flexibility, given the constant position of hip adduction in pedaling.

Hip Abduction with Bicycle Support

Slide your foot toward the rear of the bicycle.

START

Stand next to your bicycle, facing the frame. Hold the bike with both hands, one on the handlebar stem and other on the saddle. Keep your feet apart and your knees straight, and try to align your longitudinal axis on the pedal. Your elbows should remain straight to assure a distance from the bicycle that will provide adequate support and balance while performing this exercise.

TECHNIQUE

Lower your center of gravity by bending at the hips and the knee closest to the handlebars, which are your main supports. At the same time, slide the foot that is closest to the saddle toward the rear of the bicycle, keeping your knee straight and using hip abduction. Your body weight will facilitate maximum hip abduction, and thus the muscles on that side will be stretched.

LEVEL	REPS	DURATION
BEGINNER	2	20 sec
INTERMEDIATE	2	25 sec
ADVANCED	3	25 sec

CAUTION

Keep at least 14 inches (35 cm) away from the bicycle so that it acts as a supplementary support to your feet and does not obstruct the stretch or upset your balance during the performance of this exercise.

INDICATION

For all cyclists, regardless of the sport, to maintain optimum flexibility in the hip joints, because hip abduction is not part of cycling.

Upper Body Side Bend with Crossed Legs

START

Stand with the leg to be stretched crossed behind the other one, with your feet nearly in line with one another and keeping both knees straight. Your upper body should remain perpendicular to the ground. The hand corresponding to the leg being stretched can rest on your hip for greater comfort, and the other arm should remain relaxed.

TECHNIQUE

Slide the rear foot using hip adduction, so that it crosses behind the other leg, which takes on the main support function. At the same time, lean your upper body in the direction of the sliding foot and the leg being stretched, so that the hip adduction is accentuated. Even though the tension is not as evident in this stretch as in others, you probably will feel it in the lateral area of your hip.

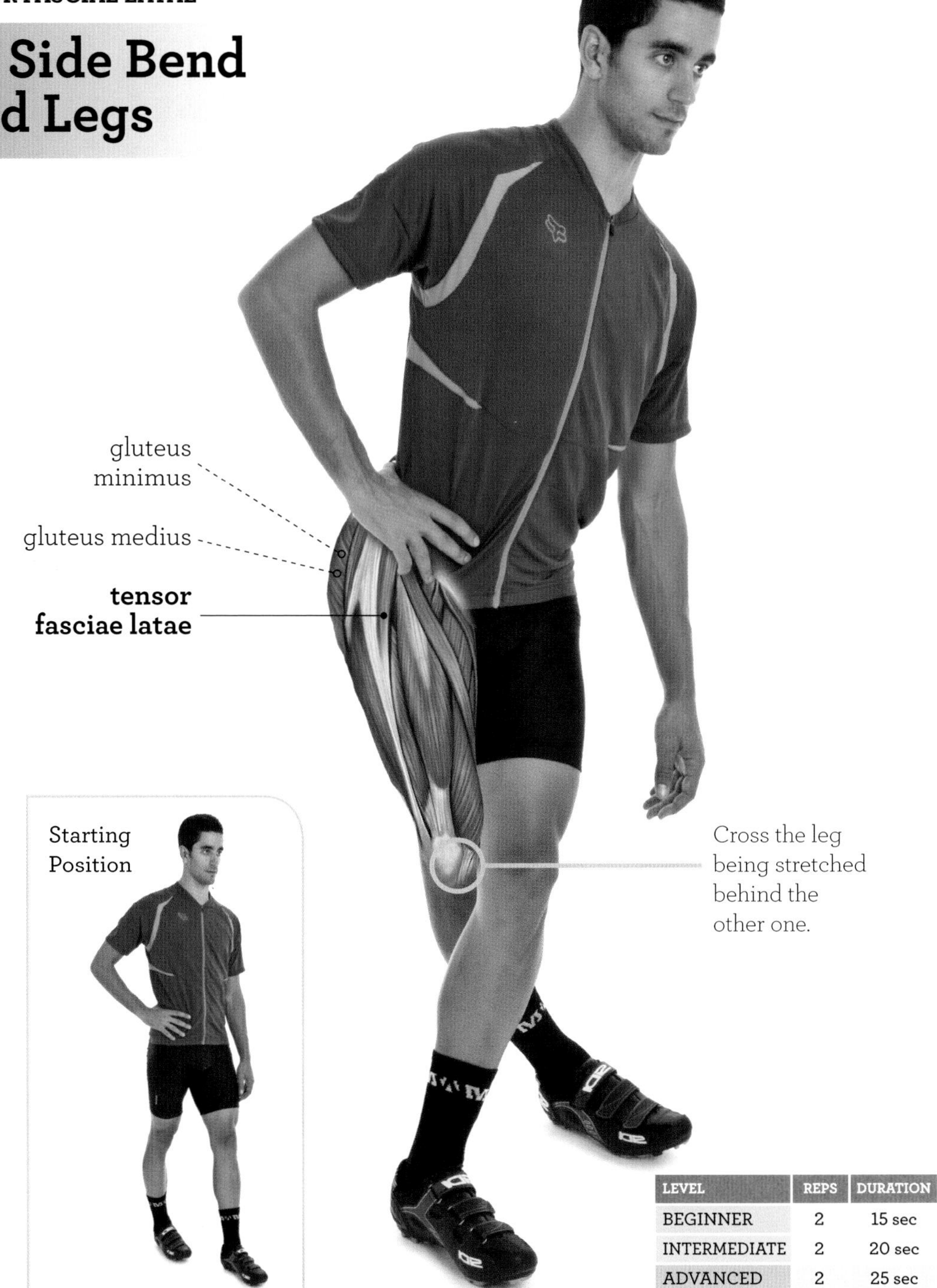

LEVEL	REPS	DURATION
BEGINNER	2	15 sec
INTERMEDIATE	2	20 sec
ADVANCED	2	25 sec

CAUTION

The major drawback to this exercise is the delicate balance during its execution, so you should do it slowly and, if possible, using a vertical support.

INDICATION

For all cyclists, especially those who do very long training rides and events, or who ride a lot, because they are more likely to develop iliotibial tract syndrome and trochanteric bursitis, which produce pain in the hip and the outside of the knee, respectively. Both afflictions can be prevented and treated with stretches for the tensor fasciae latae.

Rear Foot Cross with Support

START
Stand next to your bicycle, facing the front wheel. Hold the handlebars with both hands, squeezing the brake levers to hold the bike steady and make it a firm, unmovable support. Keep your elbows and knees straight, with your upper body bending forward very slightly.

TECHNIQUE
Slide the foot closest to the bicycle behind the other one, keeping your knee straight, while bending the hip and the knee of the other leg so that your center of gravity lowers. Your body weight will contribute to the downward motion of your body and the increased adduction of the hip being stretched, which will increase the tension on the tensor fasciae latae and facilitate the stretch.

LEVEL	REPS	DURATION
BEGINNER	2	20 sec
INTERMEDIATE	2	25 sec
ADVANCED	2	30 sec

CAUTION

As this stretch progresses, the supports on the forward foot and the bicycle will become more important, so keep the brakes applied tightly with the frame leaning slightly toward you.

INDICATION

For cyclists who ride frequently and far. This exercise will help them prevent discomfort, such as trochanteric bursitis and iliotibial tract syndrome, and can treat these if they occur. Also for bike riders in general, for maintaining the best possible overall flexibility.

Supine Chest Pull

START
Lie down on your back on the floor or a mat, with your arms along your upper body and your head. Bend both knees, and, to a lesser extent, your hips, and keep one foot on the ground. Cross the leg being stretched over the other one, resting the outer part of the ankle on the distal end of the thigh, just above the knee.

TECHNIQUE
Use both hands to grip the leg in contact with the floor just below the knee, with your fingers interlaced so that they constitute a solid grasp. Pull toward your chest with both hands without undoing the crossed legs, forcing the flexion of both hips. As the stretch advances, you will feel the tension in the deep muscles below the gluteus and in the upper back of the thigh.

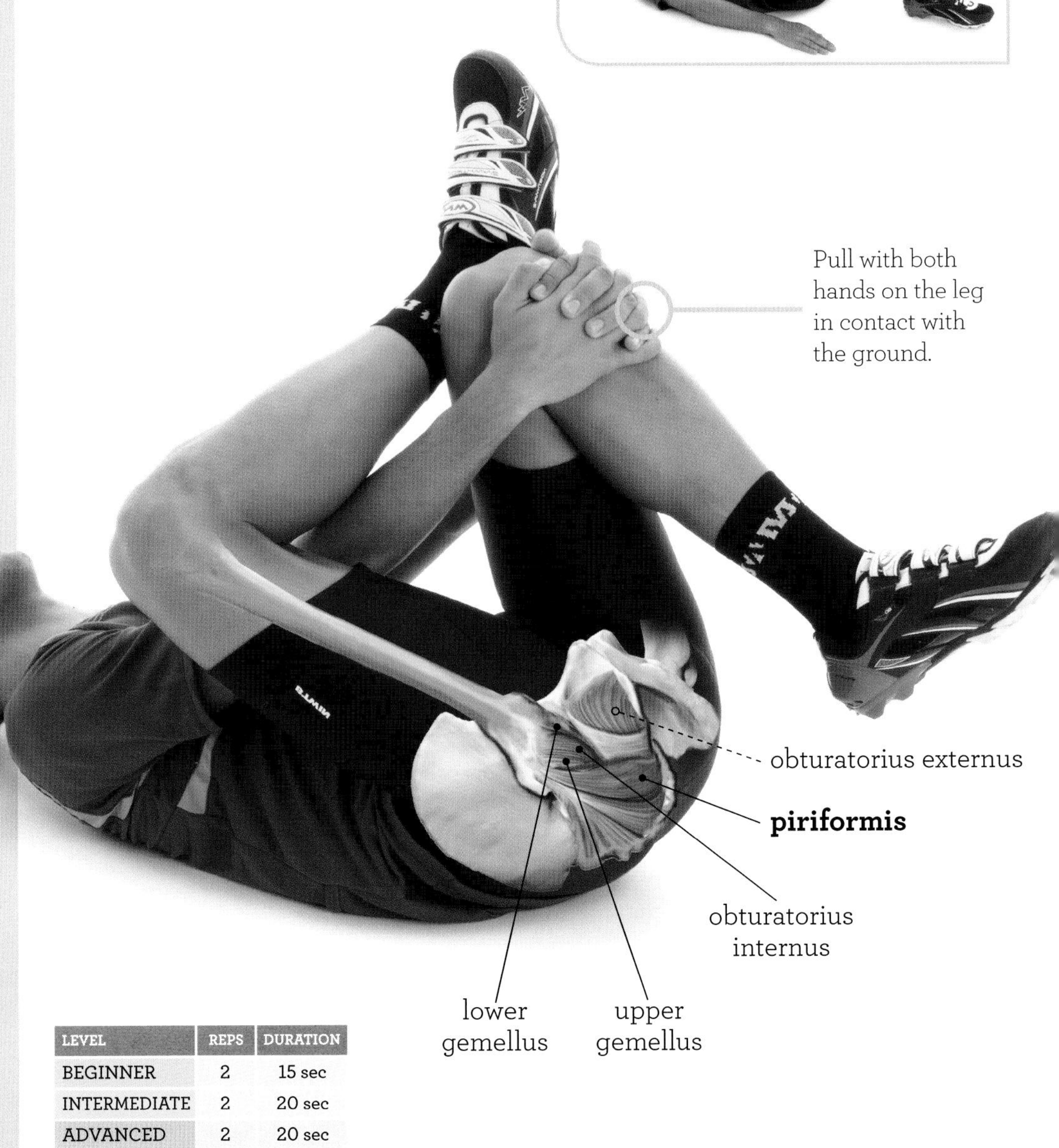

LEVEL	REPS	DURATION
BEGINNER	2	15 sec
INTERMEDIATE	2	20 sec
ADVANCED	2	20 sec

CAUTION
It's a good idea to use a mat, and to keep your head resting on the floor, with your neck relaxed, to avoid unnecessary tension in the cervical vertebrae.

INDICATION
For all cyclists, because it contributes to general flexibility. Especially for athletes who combine cycling with running, such as triathletes, because of the greater risk of pyramidal syndrome, which can involve pain in the lumbar and gluteal areas.

Cross on Bicycle

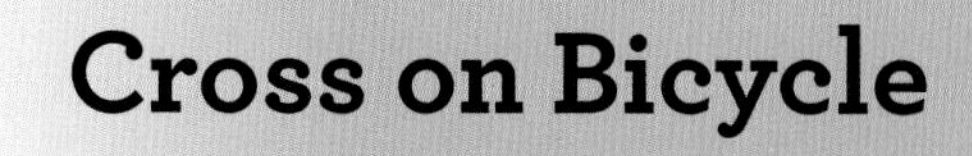

START
Stand next to your bicycle, facing the frame. Hold the saddle with one hand and the handlebars with the other, while squeezing the brake lever. Keeping one leg as your main support, raise the other one by bending your hip and knee, and rest the outside of your foot on the top tube of the frame.

TECHNIQUE
Slightly bend the knee of the support leg so that your center of gravity lowers and you increase the external rotation of the hip being stretched. As you move lower and the piriformis muscle stretches, the feeling of tension in the gluteal area and the upper back of the thigh will increase.

LEVEL	REPS	DURATION
BEGINNER	2	15 sec
INTERMEDIATE	2	20 sec
ADVANCED	2	20 sec

CAUTION

Be sure to start from a stable position, that the brakes are holding your bicycle, and that the bicycle is perpendicular to the ground or slightly leaning toward you. Also remember to avoid significantly exceeding the stretching times, or applying too much tension, due to the relatively small size of the piriformis muscle.

INDICATION

For all cyclists, because this exercise contributes to the best possible overall flexibility. Especially athletes who combine cycling and running, in order to prevent or to treat pyramidal syndrome, which can cause pain in the lumbar and gluteal regions.

Assisted Hip Extension

START
Lie down on your back on an exercise bench or some other raised support, so that your gluteus muscles are on the edge, and allow room for hip movement. Extend one leg nearly in line with your upper body, and bend the hip and knee of the other leg. Your assistant should take a position near your feet and hold both of your knees in order to perform the exercise.

TECHNIQUE
Your assistant should push with both hands, so that your flexed hip will increase its angle of flexion and your straight leg will move downward and out of alignment with your upper body. This forces greater hip extension and produces the stretch in the hip flexor muscles, as well as a feeling of tension in the groin area.

LEVEL	REPS	DURATION
BEGINNER	2	20 sec
INTERMEDIATE	2	25 sec
ADVANCED	3	25 sec

CAUTION

Make sure that the movements are done in a slow, controlled manner, and that you continually communicate with your assistant, in order to increase or stop the movement immediately if necessary.

INDICATION

For all cyclists, but especially for those who experience muscular discomfort in the groin or lumbar area due to tightness of the psoas, which commonly occurs in riders who spend a long time sitting. These symptoms may manifest themselves to a greater degree in riders who spend a lot of time training.

Hip Extension Using Bicycle Support

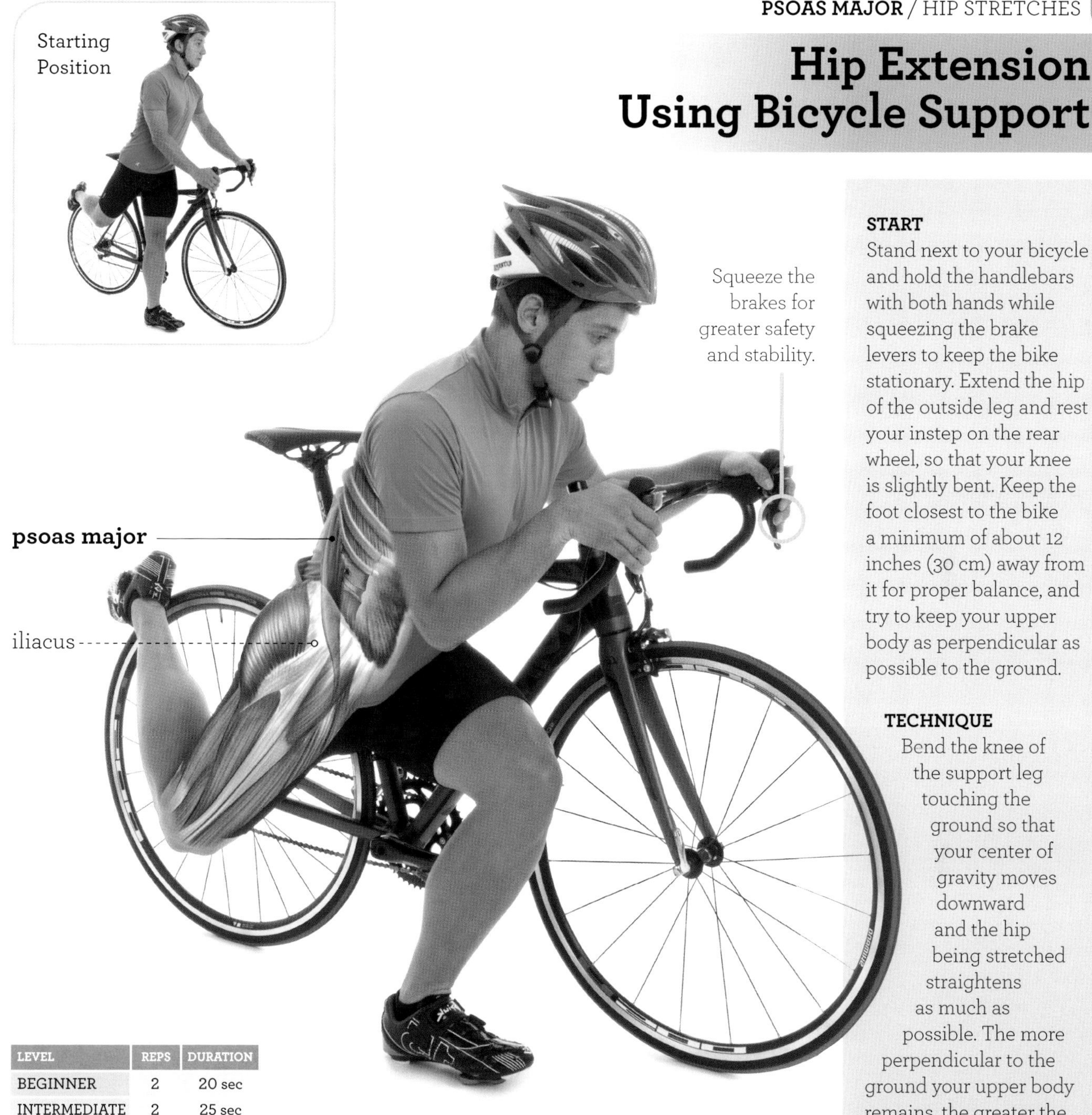

LEVEL	REPS	DURATION
BEGINNER	2	20 sec
INTERMEDIATE	2	25 sec
ADVANCED	2	25 sec

START

Stand next to your bicycle and hold the handlebars with both hands while squeezing the brake levers to keep the bike stationary. Extend the hip of the outside leg and rest your instep on the rear wheel, so that your knee is slightly bent. Keep the foot closest to the bike a minimum of about 12 inches (30 cm) away from it for proper balance, and try to keep your upper body as perpendicular as possible to the ground.

TECHNIQUE

Bend the knee of the support leg touching the ground so that your center of gravity moves downward and the hip being stretched straightens as much as possible. The more perpendicular to the ground your upper body remains, the greater the intensity of the stretch as you move downward. A feeling of tension will become evident in the groin area.

INDICATION

Before starting, make sure that the foot touching the wheel is firmly planted and that your position is stable enough to perform the exercise safely.

For cyclists in general, especially for those who train a lot or who experience tightness of the psoas, which can manifest itself as tension or muscular discomfort in the groin area or as lumbar pain and excessive anteversion of the pelvis.

Knight's Position

START
Take a position resting on a mat with one knee and one foot. Place one leg in front of you, with your hip and knee bent, each one forming about a 90° angle, and place the sole of your foot onto the mat. The thigh of the opposite leg should be in line with your upper body, with the support knee bent around 90°. Your hands can rest on your forward thigh and help sustain the position while doing the exercise.

TECHNIQUE
Move your upper body forward without changing the original supports, so that your hip and knee increase their degree of bend, while the hip and knee of the rear leg gradually increase their extension. Try to reach maximum extension in the hip being stretched and stretch your psoas major, which you will feel as tension in the groin area.

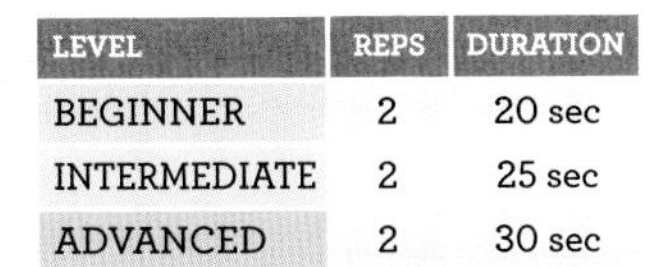

LEVEL	REPS	DURATION
BEGINNER	2	20 sec
INTERMEDIATE	2	25 sec
ADVANCED	2	30 sec

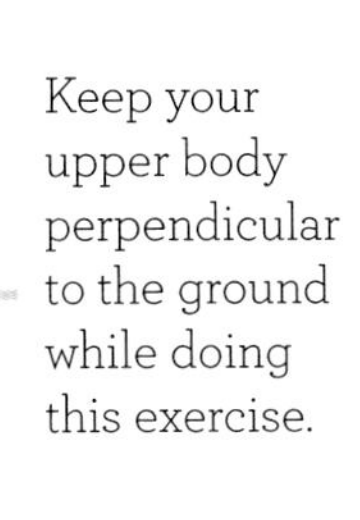

psoas major

iliacus

Starting Position

CAUTION
Try to use a padded mat. Do not do this stretch on the floor, because discomfort in the support knee would prevent you from reaching the optimum stretching point.

INDICATION
For all cyclists, because this exercise helps to maintain good flexibility in the hip flexor muscles, given their tightness in the pedaling position on the bicycle. Especially for riders who train a lot or experience discomfort in the groin or lumbar area due to tightness in the psoas.

Deep Lunge Using Bicycle Support

Starting Position

Keep your upper body perpendicular to the ground and your hips in the greatest possible extension.

psoas major

iliacus

START

Stand next to your bicycle and face forward so that your side is closest to the saddle and the frame. Move the outside leg backward one step and hold the handlebars with both hands while squeezing the brakes. Keep both knees straight.

TECHNIQUE

The next step can be done in two ways with the same results. The first one involves sliding the rear foot farther back while bending the forward knee and lowering your center of gravity. In this first option, the brakes must be applied so that the bicycle remains stationary. The second way involves keeping the rear foot in place and moving the other one forward to lengthen the lunge, so you can bend the knee and lower your center of gravity as the bicycle moves forward in concert with the movement of your body. In both cases, you should achieve maximum extension of the hip being stretched, so you should try to maintain the extension of the knee on the same side and keep your upper body perpendicular to the ground.

LEVEL	REPS	DURATION
BEGINNER	2	20 sec
INTERMEDIATE	2	25 sec
ADVANCED	2	30 sec

CAUTION

Start from a balanced position and use the option that suits you best. Always keep a distance of at least 12 inches (30 cm) between your feet and the bicycle.

INDICATION

For all cyclists, because of the relatively shortened position of the hip flexors while pedaling and the stresses to which they are subjected. Especially for cyclists who experience tightness of the psoas major, which may manifest itself as anteversion of the pelvis and tension or discomfort in the groin and lumbar areas.

Crossed Leg Pull

Starting Position

START

Sit down on the floor or a mat with one leg straight and resting on the floor, and cross the other leg over it. The crossing leg will be bent at the knee, and the foot will touch the ground beside the other knee, with the side of your ankle touching it. Turn your upper body slightly toward the side of the leg that crosses over and place your arm over it in contact with your thigh.

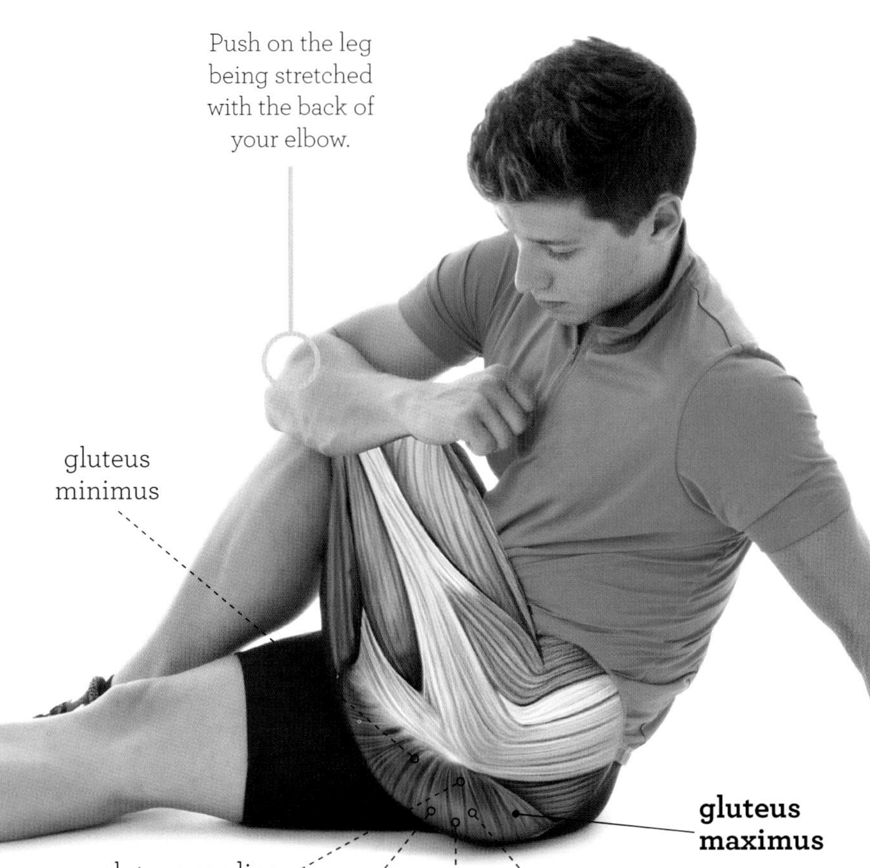

TECHNIQUE

Perform extension or retropulsion of the shoulder corresponding to the crossed arm, so that you use your elbow to push the leg being stretched and force the flexion, adduction, and internal rotation of the hip. This movement will produce a stretch in the gluteus maximus, which will manifest itself as a feeling of tension in the gluteal area.

LEVEL	REPS	DURATION
BEGINNER	2	25 sec
INTERMEDIATE	3	25 sec
ADVANCED	3	30 sec

CAUTION

This exercise involves no special risk, so you needn't take any precautions, other than making sure that your position and the pulling are kept on solid, secure supports.

INDICATION

For all cyclists, because of the great demand that the hip extensor muscles experience and the work to which they are subjected while pedaling. Especially for cyclists who ride on rough terrain and do significant climbs in training and competition. This exercise contributes to prevention and treatment of strain in the gluteus muscles.

Supine Knee Pull

LEVEL	REPS	DURATION
BEGINNER	2	25 sec
INTERMEDIATE	3	25 sec
ADVANCED	3	30 sec

START

Lie on your back on a flat surface, with your legs and arms relaxed and your head resting on the ground. Bend your hip and knee on the same side and hold the outside with the opposite hand without pulling. The other knee should remain straight and in contact with the ground.

TECHNIQUE

Pull on the bent knee, trying to cross it over your abdomen. The pull should produce the greatest possible bend in the knee and hip being stretched, as well as their adduction. If you do this exercise correctly, you will feel muscle tension in the gluteal area. Hold the pull for a few seconds before doing the stretch on the opposite side.

CAUTION

It is common to raise the head from the ground when stretching in a supine position to see how the stretch is going, but you should avoid doing this, particularly if you are doing several reps, because this could produce tension in your neck.

INDICATION

For all cyclists, because of the special demands placed on the hip extensor muscles in pedaling. Especially for cyclists who ride up long hills, take long rides, or do lots of training, because of their special exposure to strain in the hip extensors.

Knee Pull to Chest

START
Sit on the floor or a mat. Stretch one of your legs out in front of you, touching the floor. Bend the hip and knee of the opposite leg, with the corresponding foot resting on its heel. Hold the bent knee with the opposite hand and use the other one as an additional support point.

TECHNIQUE
Pull on the knee you are holding so that your heel comes off the floor and the bend in the knee and hip is accentuated. Also apply a bit of external rotation of the hip, so that the heel on the side being stretched crosses slightly over the opposite knee. Once they reach this point, you should clearly feel the muscle tension from the stretch in the gluteal area.

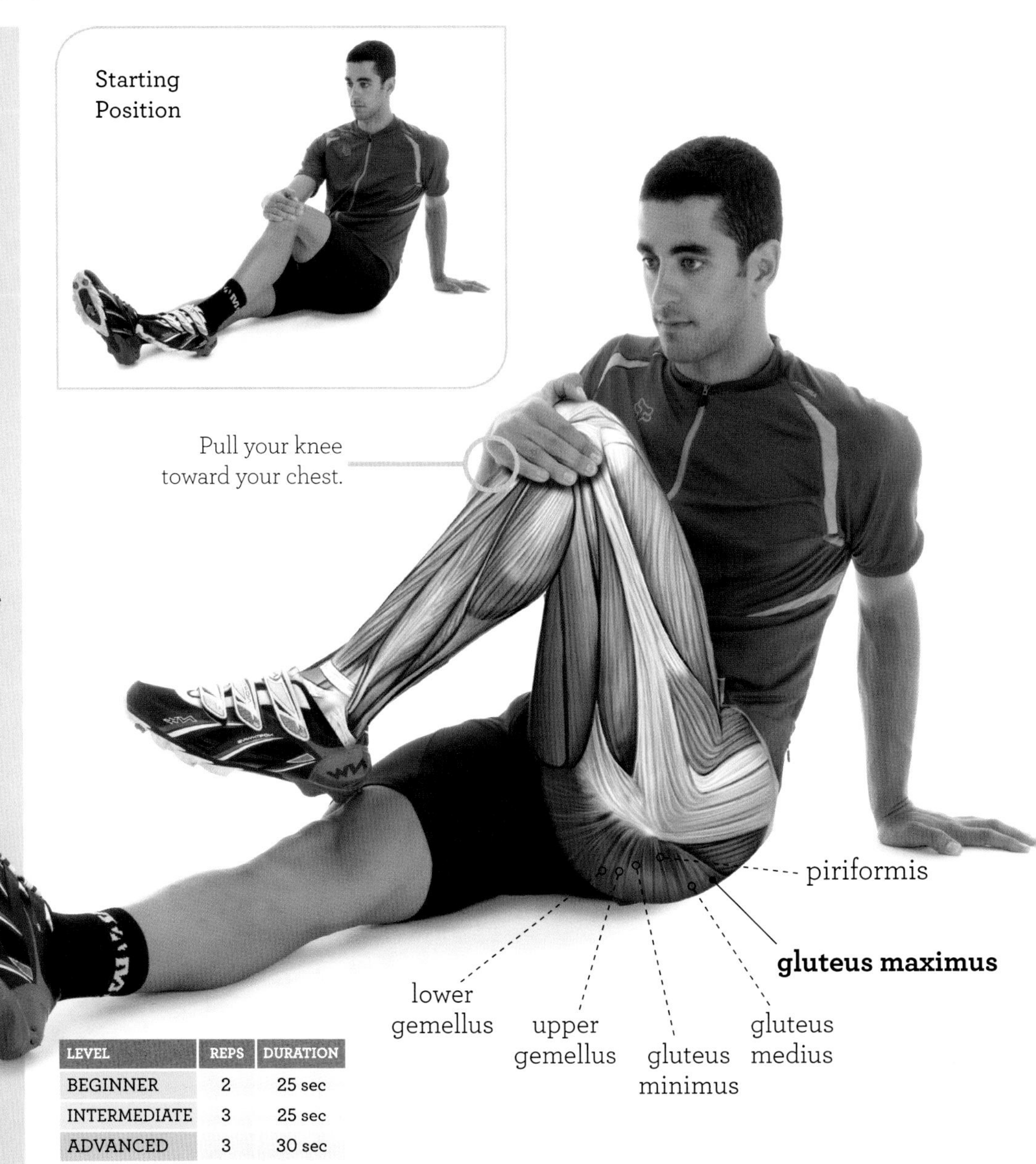

LEVEL	REPS	DURATION
BEGINNER	2	25 sec
INTERMEDIATE	3	25 sec
ADVANCED	3	30 sec

CAUTION

This exercise involves no risk, but you may need to pull on your knee with both hands to increase the pulling power if you do not feel sufficient muscle tension during the exercise to achieve the proper tension.

INDICATION

For all cyclists, because of the work of the hip extensor muscles while riding, whether in competition or for recreation. Especially for cyclists who train hard and for riders who have previously experienced muscle strain in the gluteal muscles.

Supine Pull to Chest

LEVEL	REPS	DURATION
BEGINNER	2	25 sec
INTERMEDIATE	3	25 sec
ADVANCED	3	30 sec

START

Lie down on your back with your head resting on the floor and your arms relaxed alongside your upper body. Then bend your knees and hips on both sides so that the soles of your feet touch the floor. Raise the leg being stretched and cross it over the other one, maintaining the bend in the knee.

TECHNIQUE

Lift the support foot from the floor by bending both hips and place your hands behind your thighs. Both hands should be locked together, either by interlacing your fingers or by holding one wrist with the other hand. Pull on your thighs as much as possible, trying to bring them to your chest without uncrossing your legs. The closer your thigh comes to your chest, the better the stretch. You will feel the muscle tension in the gluteal area of the crossed leg.

CAUTION

To prevent tension in your neck, remember to keep your neck relaxed and your head resting on the floor, avoiding the tendency to raise it.

INDICATION

Especially for cyclists who commonly ride long distances, train for a long time, or do rides involving long climbs or very steep hills, because of their greater exposure to muscle strain in the gluteal muscles.

Hip Flex with Bicycle Support

START

Stand next to your bicycle, facing the frame, with your feet a step apart. Use one hand to grasp the handlebars firmly, and place the other one on the saddle. Raise one foot and rest it on the top tube. You will have to bend your hip and knee. The bicycle should tilt toward you slightly. Squeeze the brake lever for added safety.

TECHNIQUE

Lean slightly toward the frame, so that you force a greater bend in the hip being stretched, and the front of your thigh comes close to or even touches your chest. This exercise provides a stretch of lower intensity for this muscle group, so the feeling of tension in the gluteal area is not as great as with other stretches.

LEVEL	REPS	DURATION
BEGINNER	1	15 sec
INTERMEDIATE	2	15 sec
ADVANCED	2	20 sec

CAUTION

This stretch is difficult, because of the delicate balance involved, so don't do it if you cannot find a stable position. Because of its low intensity, it's a good idea to combine it with other stretches for the hip extensor muscles.

INDICATION

For cyclists who do lots of training, who like to ride long distances, or who ride routes involving long, steep climbs, because of the special requirements on the hip extensor muscles. Also for cyclists who repeatedly experience muscle strain in the gluteal muscles.

Supine Unilateral Pull to Chest

Starting Position

Use both hands to pull your thigh toward your chest.

gluteus medius

gluteus maximus

gluteus minimus

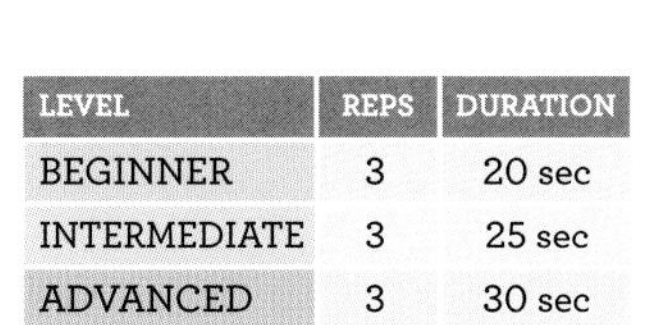

LEVEL	REPS	DURATION
BEGINNER	3	20 sec
INTERMEDIATE	3	25 sec
ADVANCED	3	30 sec

START

Lie down on your back on a flat surface and bend your hip and knee on the same side while keeping the opposite hip and knee straight and in contact with the floor. Use both hands to hold the raised knee, and keep your neck relaxed and your head on the floor.

TECHNIQUE

Pull the raised knee toward your chest, forcing the bend in the hip being stretched, and moving the front of your thigh to your chest, while keeping the other leg in its original position and the rear of your hip in contact with the floor. As in the previous case, this exercise produces less intensity than the other suggestions for the gluteal muscles, so the feeling of muscle tension may be very slight.

CAUTION

This exercise requires no particular caution beyond observing the recommendation of keeping your head on the floor and your neck relaxed to avoid tension in that area.

INDICATION

For all cyclists, especially those who ride a lot or who ride routes with long, steep hills, because of the strain this can produce in the hip extensor muscles.

LOWER LIMB STRETCHES

Cyclists' legs are probably their most prominent feature, and this has little to do with genetics or whims of nature. Rather, it is the result of the hard work to which the muscles of this part of the body are subjected. Pedaling requires, of course, work from the quadriceps femoris to straighten the knee and push the pedal forward and downward with sufficient power to move the bike forward, especially on long climbs or on rising terrain. Also, even though the ischiotibial muscles are used less intensively, they contribute to the forward thrust and to raising the pedals by helping to bend the knee. These efforts, under conditions of significant exhaustion, can lead to problems, such as tendonitis of the kneecap, tendonitis of the quadriceps, synovial plica syndrome, tendonitis of the pes anserinus, and tendonitis of the biceps femoris, in addition to the usual muscle strains. Most of these problems are due to the repeated exertions and movements that are common in cycling, and they can cause pain in different parts of the knee and surrounding areas. At the same time, strengthening the muscles involved and stretching can prevent or reduce these problems, resulting in healthier and more comfortable cycling activity.

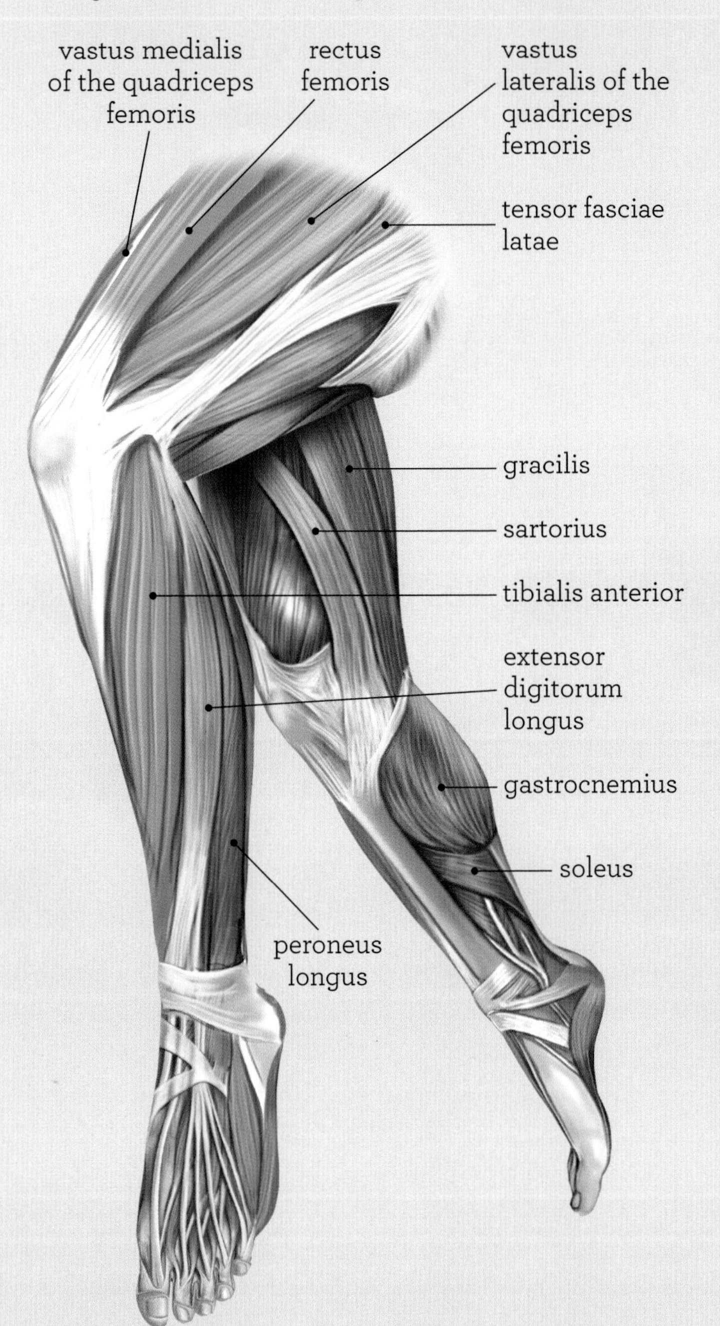

In addition, the muscles responsible for plantar flexion of the ankle also exert great effort in advancing and driving the foot while pedaling with the distal part of the metatarsals in this phase. Thus, this thrust, which requires keeping the front of the foot lower than the heel, contributes to the pedaling cycle and the forward motion of the bicycle, and, just as with the quadriceps, this work is particularly intense during climbs. As a result of this effort, problems may arise, such as tendonitis of the Achilles tendon, tendonitis of the tibialis posterior, and strain in the calf and the soleus, which can manifest themselves as pain in the calf and the rear of the ankle. Stretches can contribute to prevention and treatment of these discomforts.

QUADRICEPS FEMORIS

This muscle is the main one involved in straightening the knee. It is made up of the following four muscles.

Rectus femoris: This muscle originates at the anteroinferior iliac spine and at the roof of the acetabulum, and it inserts at the tibial tuberosity through the tendon of the quadriceps femoris, the kneecap, and the patellar ligament. It shares an insertion with the vastus medialis, lateralis, and intermedius muscles. Its main function is extension of the knee, although it is also used in bending the hip.

Vastus lateralis: This muscle originates at the trochanter magnus and the lateral labium of the femoral linea aspera, and it shares functions with the vastus medialis and intermedius in straightening the knee.

Vastus medialis: This muscle originates at the middle labium of the linea aspera and the intertrochanteric line of the femur.

Vastus intermedius: This muscle originates at the anterolateral surface of the body of the femur.

BICEPS FEMORIS
This is considered one of the ischiotibial muscles, even though this applies only to its long head. It originates at the sciatic tuberosity, whereas its short head originates at the lateral labium of the linea aspera of the femur. It inserts at the head of the fibula, and its main function is bending the knee, although its long head also contributes to hip extension.

SEMITENDINOSUS
This ischiotibial muscle originates at the sciatic tuberosity, and it inserts at the middle edge of the tibial tuberosity, the pes anserinus. Its functions, which it shares with the semimembranosus muscle, are bending the knee and straightening the hip.

SEMIMEMBRANOSUS
This muscle shares its origin with the semitendinosus, and it inserts at the middle condyle of the tibia.

GASTROCNEMIUS
This muscle originates above the lateral and middle epicondyles of the femur, and it inserts at the tuberosity of the calcaneus or heel bone, through the Achilles, or heel bone tendon. Its main function is plantar flexion of the ankle, although it is used in bending the knee.

SOLEUS
This muscle originates at the posterior face of the head and neck of the peroneus and the soleus line of the tibia, and it shares an insertion with the gastrocnemius. Its function is plantar flexion of the ankle.

TIBIALIS POSTERIOR
This muscle originates at the interosseous membrane and the posterior part of the tibia and the peroneus. It inserts at the base of metatarsals two through four and in the cuneiform and navicular bones. Its main function is plantar flexion of the ankle, although it is used in ankle inversion.

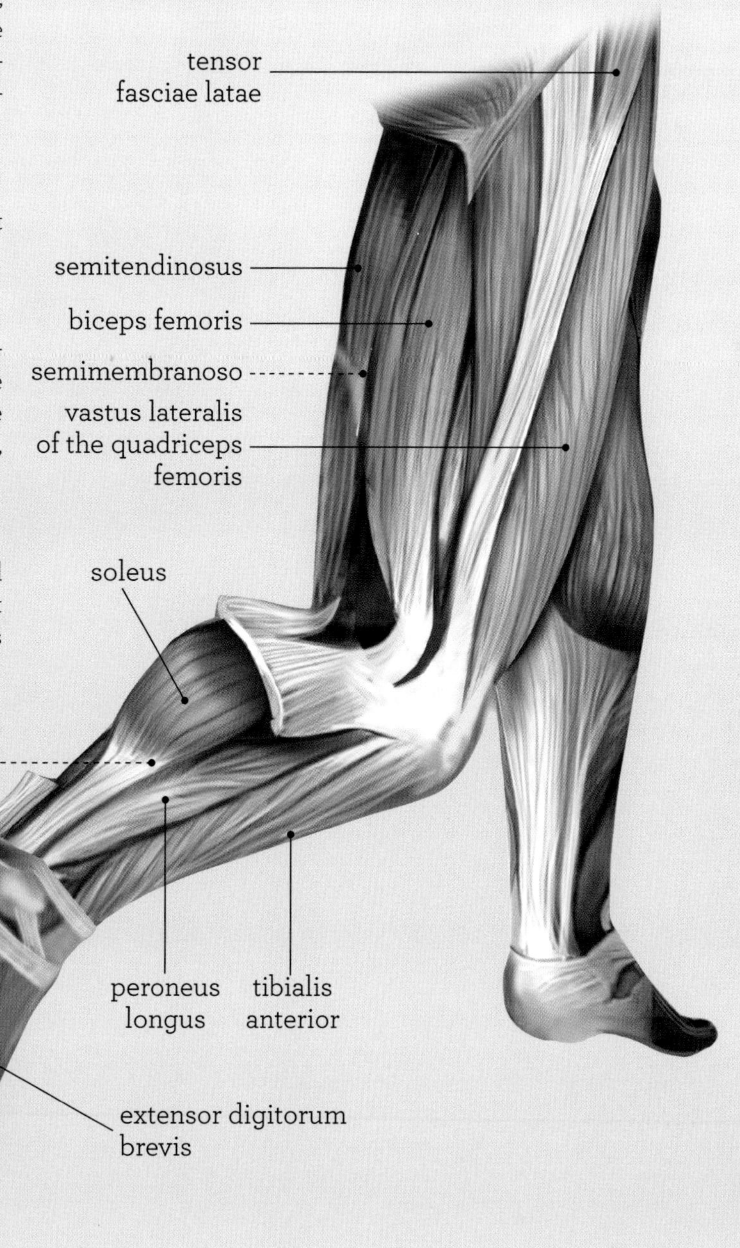

Assisted Knee and Hip Extension

START
Lie face down on a flat surface and place your hands beneath your chin to serve as a headrest. Keep your legs straight and fairly close together. Your assistant should take a position next to you, kneeling by your thighs. Then your assistant should hold your farthest leg by the front of your thigh and knee, and should raise your foot until your knee is bent at a 90° angle.

TECHNIQUE
Next, your assistant should pull on your ankle and lift your knee from the floor, bending it and extending your hip as far as possible. Soon, you will feel the muscle tension in the front of your thigh as a result of the stretch in the quadriceps femoris. If you want to increase the intensity, you can increase the distance between the knee and the floor.

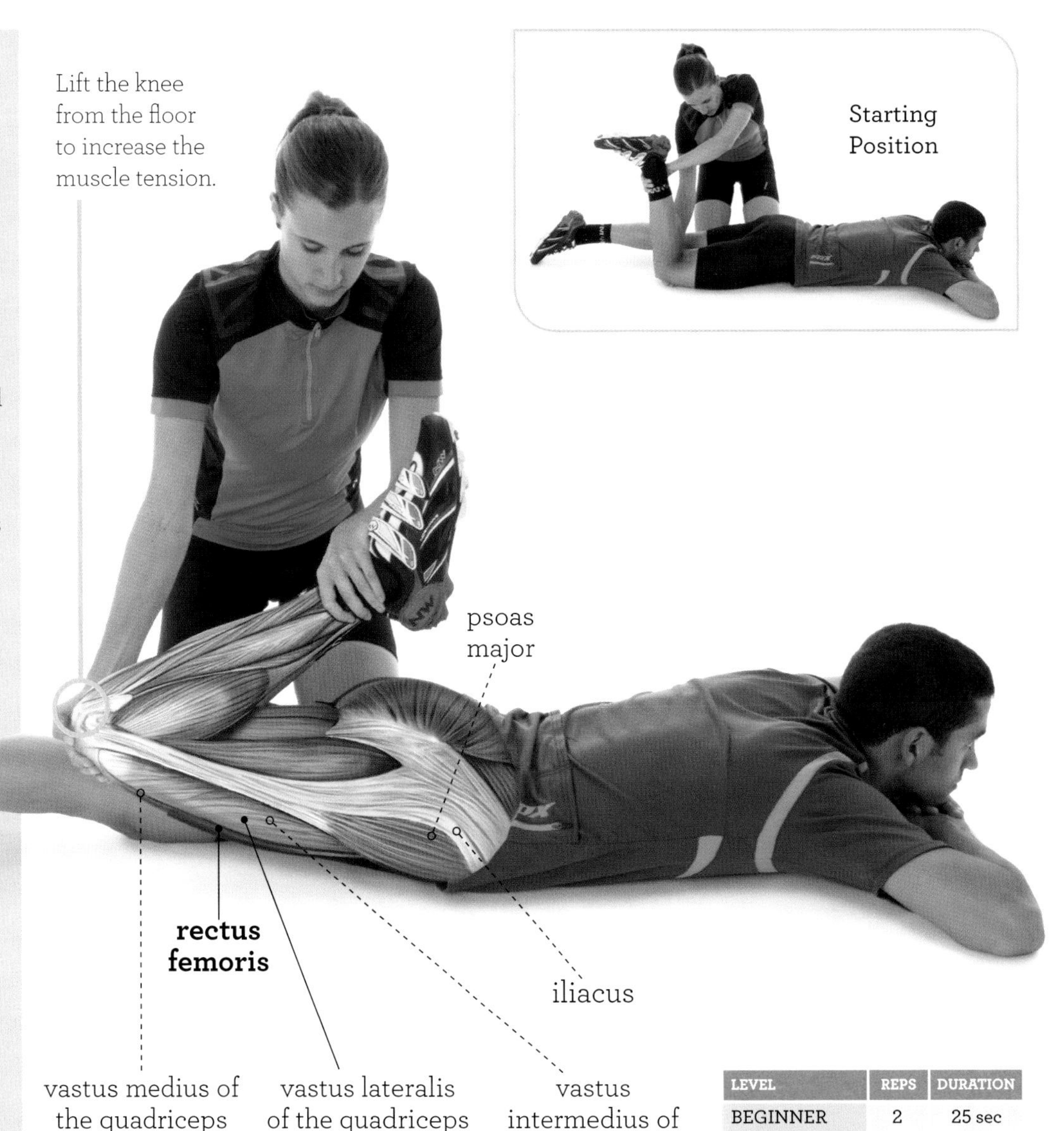

LEVEL	REPS	DURATION
BEGINNER	2	25 sec
INTERMEDIATE	3	25 sec
ADVANCED	3	30 sec

CAUTION

As with all exercises involving the help of an assitant, this exercise should be done slowly, with constant communication to assure that the stretch does not go beyond the threshold of pain and that there is no threat to the muscles involved.

INDICATION

For all cyclists, because of the work performed by the knee extensor muscles in pedaling, especially in pushing the pedal forward and downward. Also for cyclists who experience muscle tension in the front part of the thigh, as well as for those who want to prevent or treat tendonitis of the quadriceps and knee tendons in nonacute phases.

Flamingo Position

Starting Position

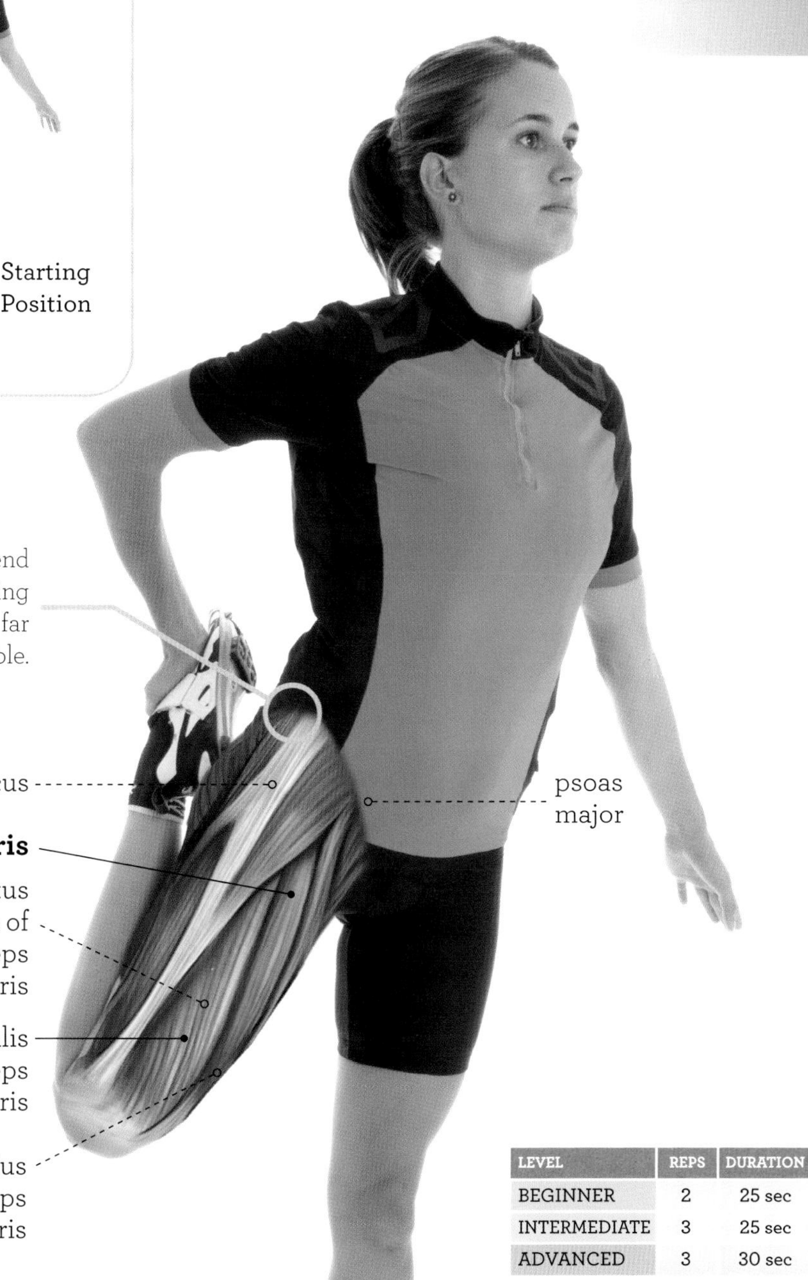

START

Stand and raise one foot by bending your knee and, to a lesser degree, your hip. Using the hand on the same side, hold the front of your ankle to hold your leg up. You can do this exercise without support, or you can use a vertical support if you need more stability or if you are going to do the exercise in long sets.

TECHNIQUE

Pull the ankle upward so that you maintain the maximum knee bend and the hip extension increases considerably. As the hip extension increases, the muscle tension caused by the stretch will make itself felt in the front of the thigh.

LEVEL	REPS	DURATION
BEGINNER	2	25 sec
INTERMEDIATE	3	25 sec
ADVANCED	3	30 sec

CAUTION

This exercise involves no specific risk, but it's a good idea to use a vertical support or to do it facing an assistant to improve your balance, especially if you do long sets.

INDICATION

For all cyclists, because of the major involvement of the knee extensor muscles in pedaling. Especially for those who regularly do long rides, climb long, steep hills, feel muscle tension in the front of the thigh, or tend to experience discomfort in the quadriceps and patellar tendons.

Knight's Position with Pull

START

Assume the knight's position on a mat—that is, resting on one knee and one foot, keeping both knees bent around 90°. The thigh corresponding to the support knee will be perpendicular to the floor. Rotate your upper body toward the side of the rearmost foot, which is the one being stretched, and look at it. Move your arm rearward toward your ankle to facilitate performance of the exercise.

TECHNIQUE

Increase the bend in the rear knee to lift it from the floor, and hold your foot by the instep, using the hand on the same side. Pull on your instep to increase the bend in the knee, while leaning forward slightly with your upper body, thus increasing the hip extension on the hip being stretched and the bend in the opposite one. As your position moves forward, the degree of stretch will increase and you will feel the muscle tension in the front of your thigh.

LEVEL	REPS	DURATION
BEGINNER	3	20 sec
INTERMEDIATE	3	25 sec
ADVANCED	3	30 sec

CAUTION

Use a mat or other cushion under your knee in this exercise and in others that require support on this joint. Remember to combine the ankle pull with the hip extension to achieve the greatest degree of stretch.

INDICATION

For cyclists who feel muscle tension in the front of the thigh, because of intense work by the knee extensor muscles, which may occur to a greater degree in riders who do their usual rides or training sessions on stretches with long or very steep climbs.

Unilateral on Side

LEVEL	REPS	DURATION
BEGINNER	2	25 sec
INTERMEDIATE	2	30 sec
ADVANCED	3	30 sec

START

Lie down on your side on a flat surface. You can use a mat. Bend the knee and hip of your top leg, the one being stretched, so that it is tucked up. Using the hand on the same side, hold your foot by the instep so that your ankle is in plantar flexion.

TECHNIQUE

Pull rearward on your foot, maintaining maximum knee bend and causing hip extension, as if you were trying to touch your gluteus with the sole of your foot. You will feel the tension in the front of your thigh, probably with the greatest intensity in the proximal third. The other leg should remain in its original position.

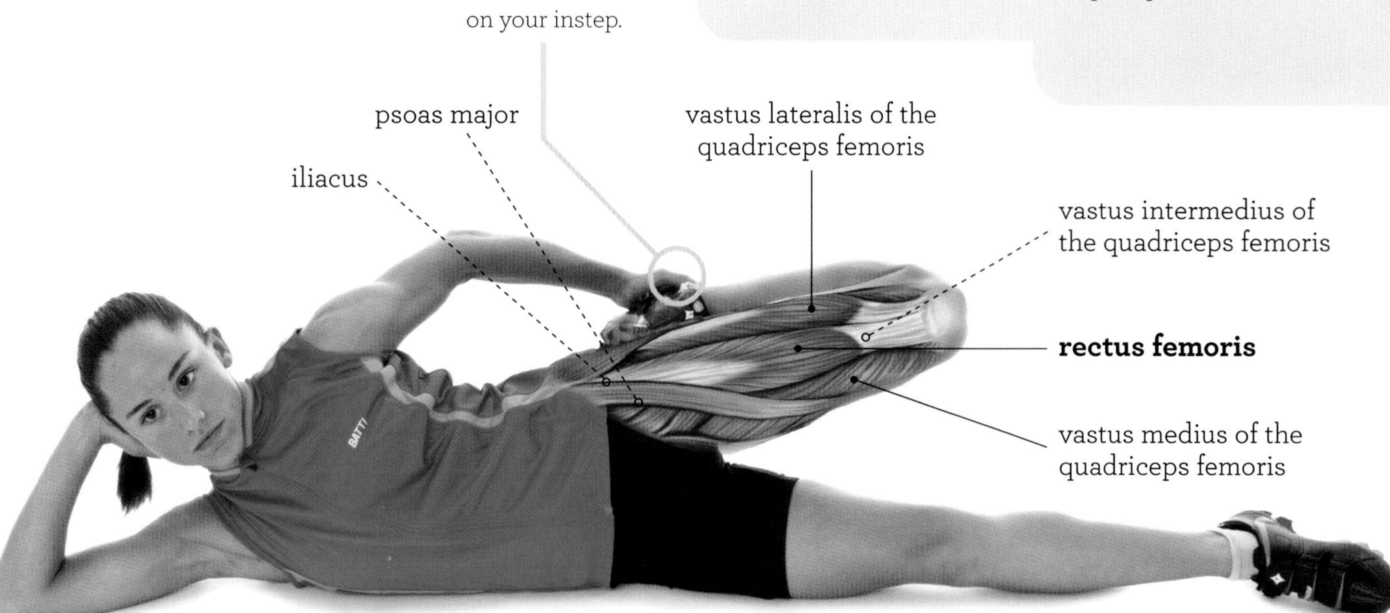

CAUTION

Avoid arching your back or resorting to anteversion of the pelvis to expand the range of the exercise, because that will increase neither the intensity nor the effectiveness of the stretch and could cause discomfort, especially in the lumbar vertebrae.

INDICATION

For cyclists who experience muscle tension in the front of the thigh, who tend to experience discomfort in the quadriceps and patellar tendons, who do mainly long rides in training, or who ride hard up long, steep climbs.

Standing with Rear Bicycle Support

START
Stand with your back to the bicycle. Hold it with both hands, one on the saddle and the other on the handlebars, squeezing one brake lever to keep the bike from moving while it supports you. Your feet should be at least 12 inches (30 cm) away from the frame, which should tilt slightly toward you. Bend one knee and support the instep of your foot on the top tube.

TECHNIQUE
Bend the support knee so that your center of gravity moves downward. This will increase the knee bend and the hip extension on the side being stretched. As you move downward, and while keeping your foot anchored on the frame, you will stretch the quadriceps femoris and feel tension in the front of your thigh, especially in its proximal third.

LEVEL	REPS	DURATION
BEGINNER	1	15 sec
INTERMEDIATE	2	15 sec
ADVANCED	2	20 sec

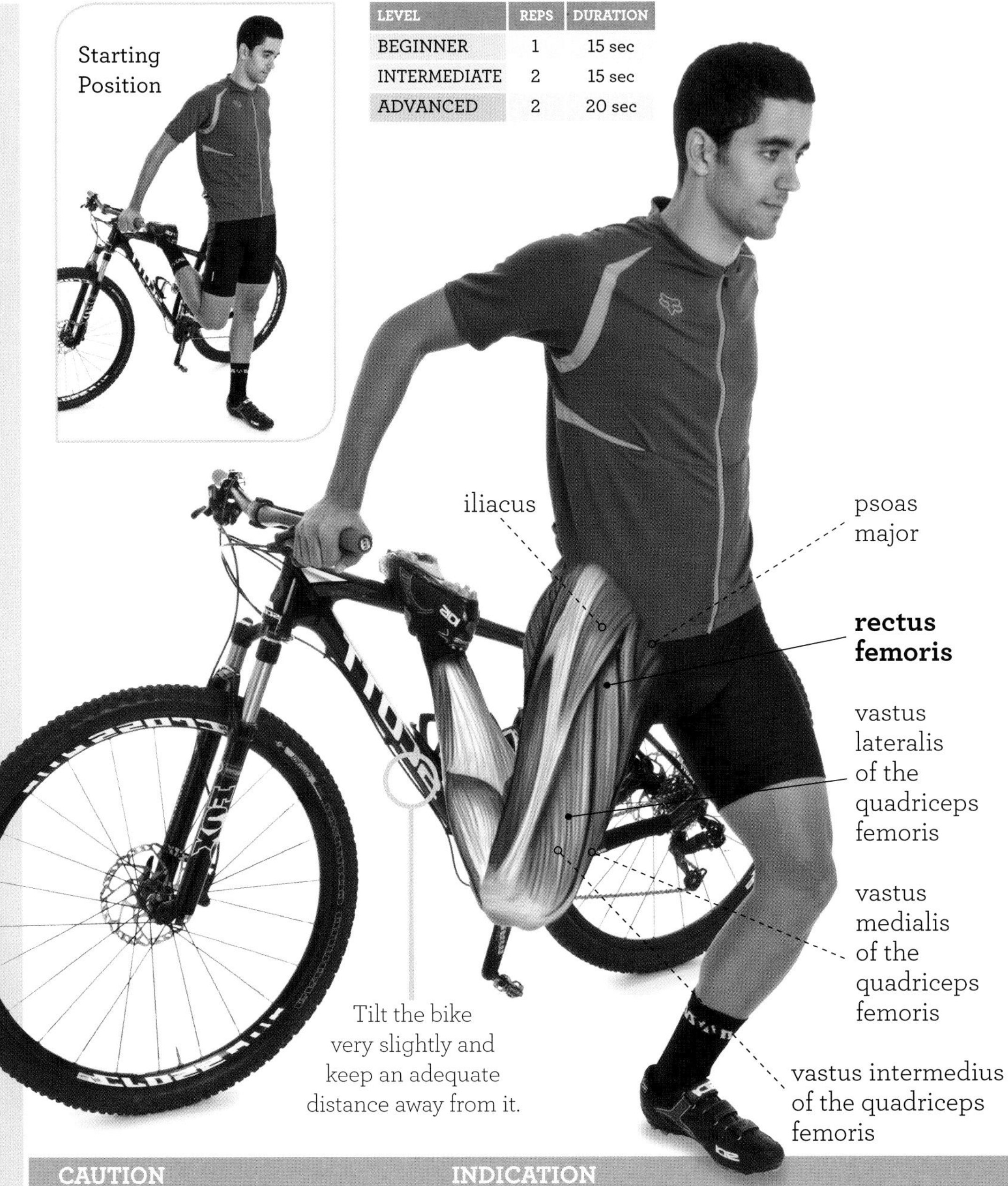

CAUTION

Keep an adequate distance from the bicycle to maintain stability, even though this involves a slight reduction in the intensity of the stretch. In this case, as in others, safety is more important than increasing the intensity.

INDICATION

For all cyclists, because of the special involvement of the knee extensor muscles in pedaling. Especially for cyclists who experience tension in the front of the thigh and feel discomfort in the quadriceps and patellar tendons and for hill climbers and time racers.

Bilateral on Knees

LEVEL	REPS	DURATION
BEGINNER	1	15 sec
INTERMEDIATE	2	15 sec
ADVANCED	2	20 sec

START
Kneel down with your ankles in plantar flexion, and move downward until the back of your thighs and your calves touch one another and you remain sitting on the inner part of your heels. Lean backward and rest your palms on the floor so that your arms are perpendicular to it, with your elbows straight.

TECHNIQUE
Slide the palms of your hands forward and keep leaning your upper body rearward as if you were trying to lie down. Your hips will increase their extension as your back approaches the floor and your knees will remain in total flexion. You can also keep your hands in their original position and bend your elbows to encourage the backward lean, especially if your level of flexibility does not allow you to lie back far enough to touch the floor.

psoas major

iliacus

rectus femoris

Extend your hips and keep your knees bent as far as possible.

vastus intermedius of the quadriceps femoris

vastus lateralis of the quadriceps femoris

vastus medius of the quadriceps femoris

CAUTION
Stop the exercise if you feel discomfort in your ankles or back, and go only as far as you feel comfortable, avoiding pain. Use a mat whenever possible.

INDICATION
For hill climbers, for time racers, for cyclists who experience muscle tension in the front of the thighs, and for frequent riders, because of the special demands that these activities place on the knee extensor muscles.

Bilateral in Inverted V-position

START

Stand with your feet a short distance apart. Your lower limbs, upper body, and neck should be perpendicular to the floor, with your hands resting on the upper sides of your thighs.

TECHNIQUE

Bend your hips to try to touch the floor with the palms of your hands, right in front of your toes. Your knees should remain totally straight, and you will have to focus on anteversion of your pelvis more than bending the upper body, because the former will be the determining factor for achieving the proper intensity of the stretch. As your hands approach the floor, you will feel the tension from the stretch in the back of your thighs and knees.

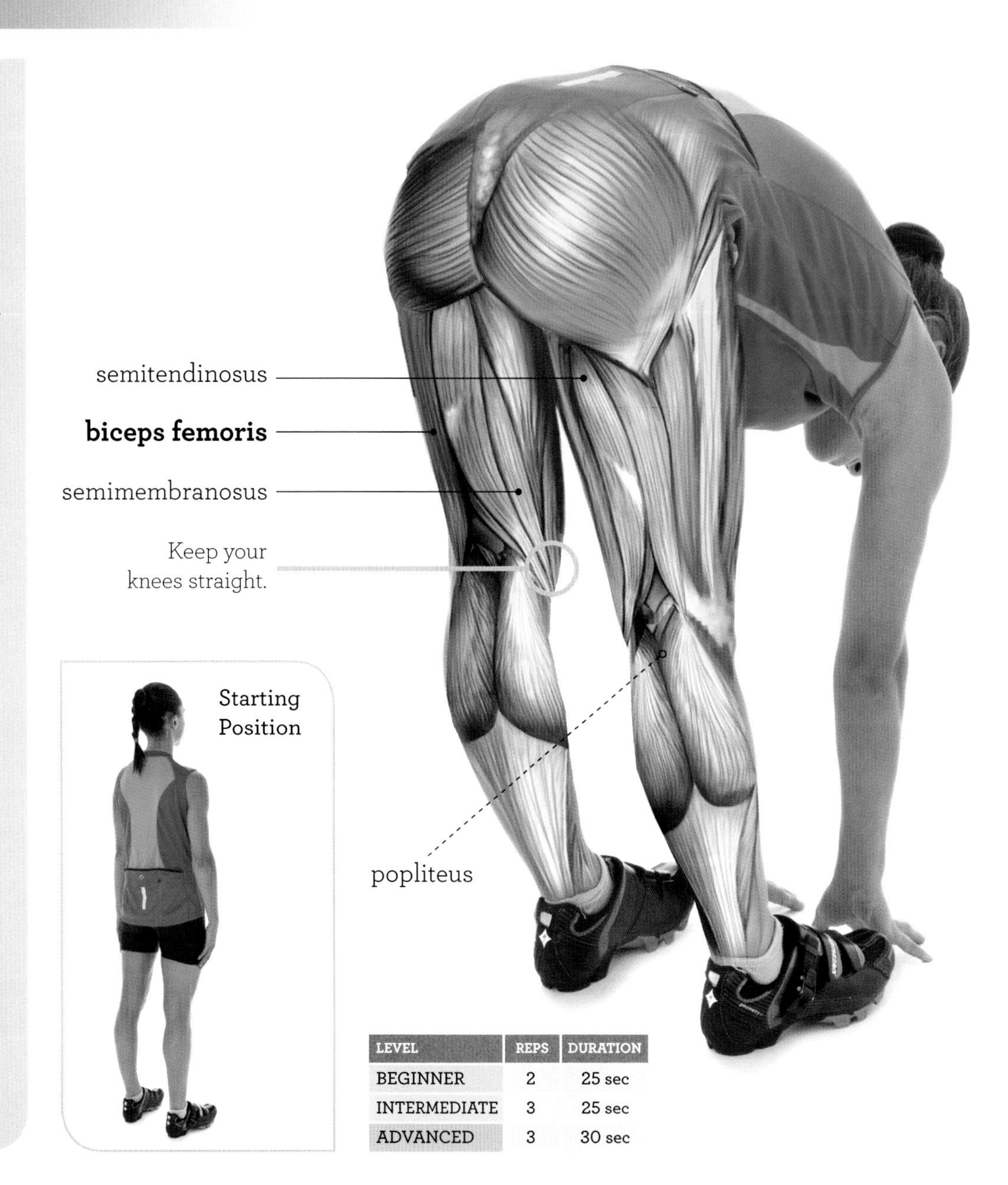

LEVEL	REPS	DURATION
BEGINNER	2	25 sec
INTERMEDIATE	3	25 sec
ADVANCED	3	30 sec

CAUTION

To the extent possible, avoid bending your upper body, because this undermines the stretch and can cause discomfort, especially in the lumbar vertebrae. However, remember that it is not essential to eliminate this bend completely.

INDICATION

For cyclists who experience tension or strain in the back of the thighs, who feel discomfort in the inner or outer face of the knee originating at the insertion of the ischiotibials, who train a tremendous amount, or who race very long distances. Also for cyclists in general, because of the involvement of the knee flexor muscles in pulling rearward and upward in the pedaling motion.

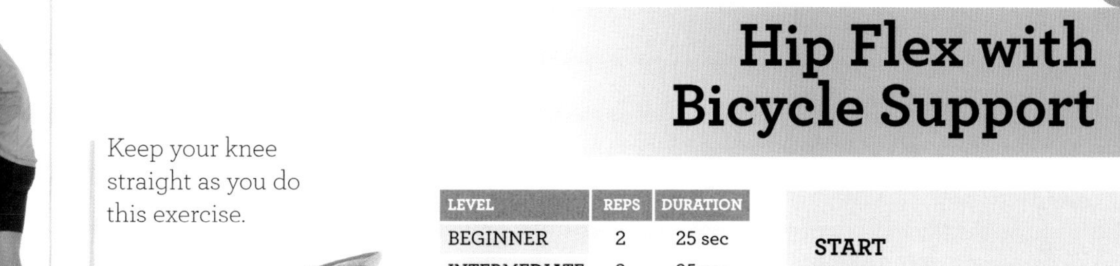

Hip Flex with Bicycle Support

Starting Position

Keep your knee straight as you do this exercise.

LEVEL	REPS	DURATION
BEGINNER	2	25 sec
INTERMEDIATE	2	25 sec
ADVANCED	2	30 sec

popliteus

biceps femoris

semitendinosus

semimembranosus

START

Take a position facing the frame, about one step away from your bicycle. Hold the handlebars with one hand, while squeezing the brake lever to keep the bicycle stationary. Raise the leg closest to the rear wheel and rest it on the saddle so that the rear of your ankle touches it.

TECHNIQUE

Straighten the knee being stretched and try to reach and hold the raised foot with the hand on the same side. You will need to lean your upper body forward and bend it slightly. Increase the hip bend and keep your knee straight. This will produce the best stretch in the ischiotibial muscles.
You will quickly feel the tension in the rear of your thigh, and it will be easily detectable.

CAUTION

Even though you will have to bend your upper body to reach your foot, this is not the main purpose of the exercise, so stress the anteversion of the pelvis. This is what will determine the usefulness of the exercise.

INDICATION

For all cyclists, because of the involvement of the ischiotibials in pedaling and their importance in general flexibility and correct posture, including on the bicycle. Especially for cyclists who experience muscle tension in the rear of the thigh or discomfort in the inner or outer faces of the knee, originating from the activity of the ischiotibials.

Seated Unilateral Hip Flex

Starting Position

LEVEL	REPS	DURATION
BEGINNER	2	20 sec
INTERMEDIATE	2	25 sec
ADVANCED	2	30 sec

START
Sit down with one of your legs stretched out in front of you and the other one rotated inward, with hip abduction and your knee bent, so that it is drawn up. Place your hands on the straight knee and keep your upper body perpendicular to the floor. Look at your forward foot.

TECHNIQUE
Slide your hands forward and try to reach your foot and hold it so that your fingers go beyond your toes and touch the front part of your sole. This will produce a slight additional pull. Try to achieve the longest range of movement using hip anteversion, even though a certain amount of upper body bend will be necessary. You will feel the tension in the back of your thigh and knee.

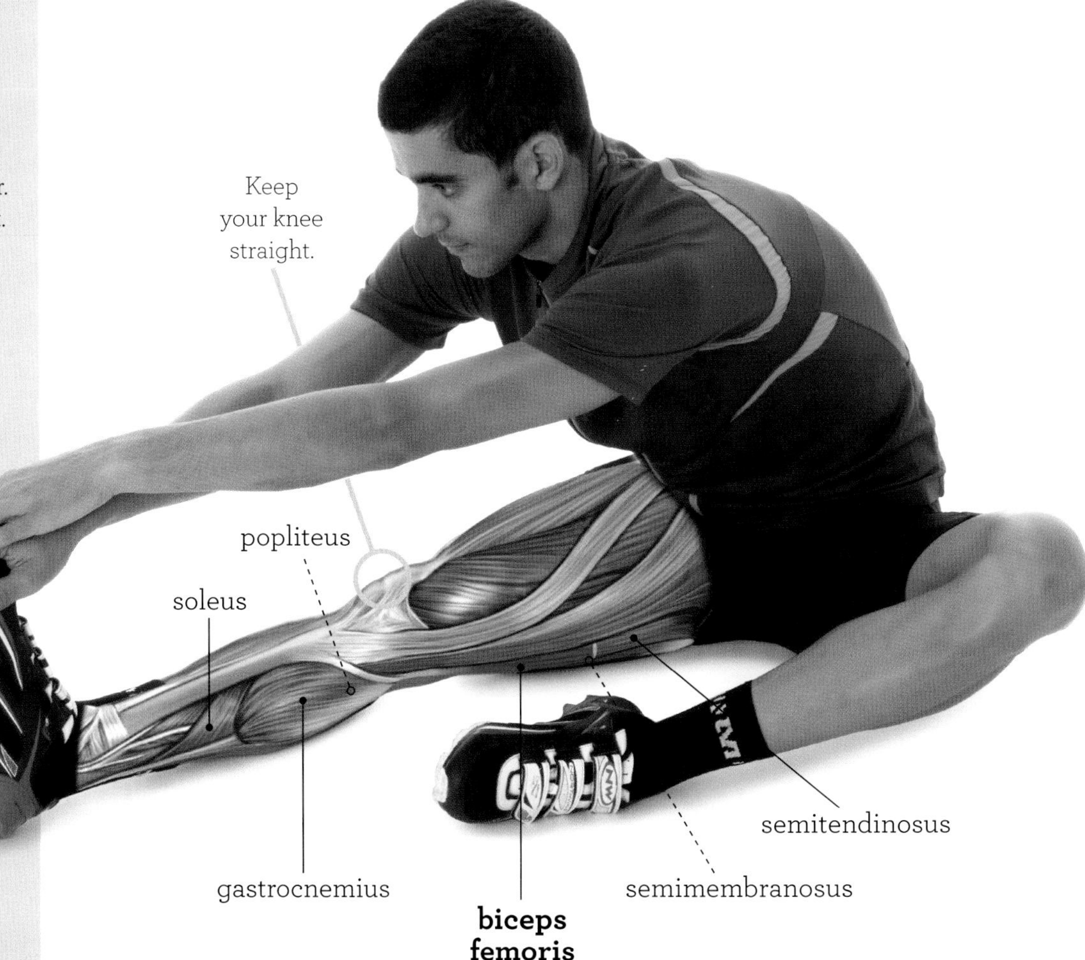

CAUTION
It's a good idea to reduce the bend in your lumbar vertebrae and use anteversion of the pelvis, both to increase the intensity of the stretch and to protect your lumbar vertebrae.

INDICATION
For cyclists who experience muscle tension in the back of the thigh, who arch their backs excessively while riding, or who have improper posture, because of retroversion of the pelvis. Also for cyclists in general, because of the involvement of the ischiotibials in pedaling, in order to maintain adequate overall flexibility.

Supine with Raised Leg

Starting Position

START

Lie down on your back and straighten one of your legs, keeping the other one drawn up. Keep your head on the floor and relax your neck. Hold the calf closest to your upper body with both hands so that your thumbs face forward and your four fingers are on your calf.

TECHNIQUE

Straighten the knee that you are holding, without letting go. At this instant, you will feel considerable muscle tension in the back of your thigh and knee. To increase the intensity of the stretch, you can pull on your leg and bring it a little closer to your chest, but this will be necessary only for cyclists who have greater flexibility.

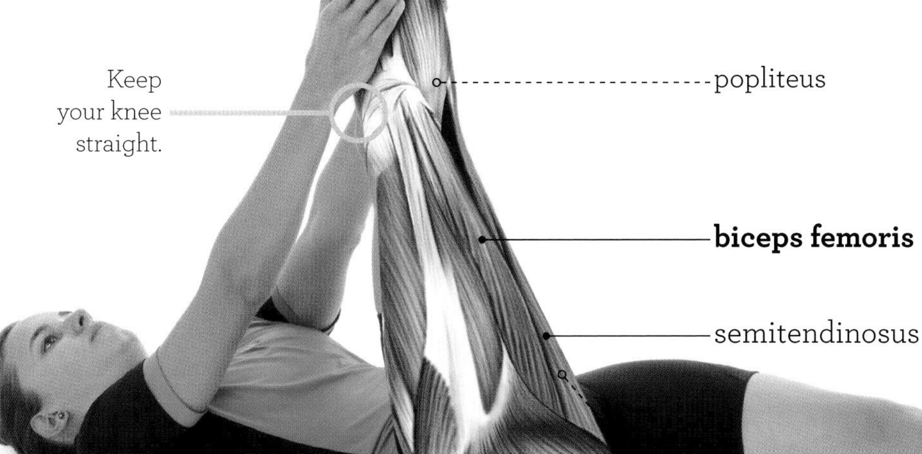

LEVEL	REPS	DURATION
BEGINNER	2	20 sec
INTERMEDIATE	2	25 sec
ADVANCED	2	30 sec

CAUTION

Try to keep your head on the floor to avoid tension in the neck while doing the exercise. This also applies to the other exercises that start form a supine position.

INDICATION

For cyclists who experience muscle tension in the back of the thigh or discomfort in the inner or outer face of the knee, or who bend their spines excessively when on the bicycle. Also for cyclists who want to maintain the best possible overall flexibility and for riders with improper posture due to retroversion of the pelvis.

Assisted Dorsal Ankle Flex

START
Sid down with one leg straight and the other one drawn up so that your hip and knee are bent and the sole of your foot touches the floor. Lean slightly rearward with your upper body and rest both hands on the floor. Your assistant should take a position by your front foot and should hold it by the toes and the heel, keeping your ankle in plantar flexion or a neutral position, while lifting if off the floor a bit.

TECHNIQUE
Your assistant should hold your heel tightly to keep it stationary and push on your toes to put the ankle into a position of maximum dorsal flexion, while your knee remains straight. You will feel the tension in your calf and the back of your knee, which indicates the effectiveness of the stretch.

LEVEL	REPS	DURATION
BEGINNER	2	20 sec
INTERMEDIATE	2	25 sec
ADVANCED	3	25 sec

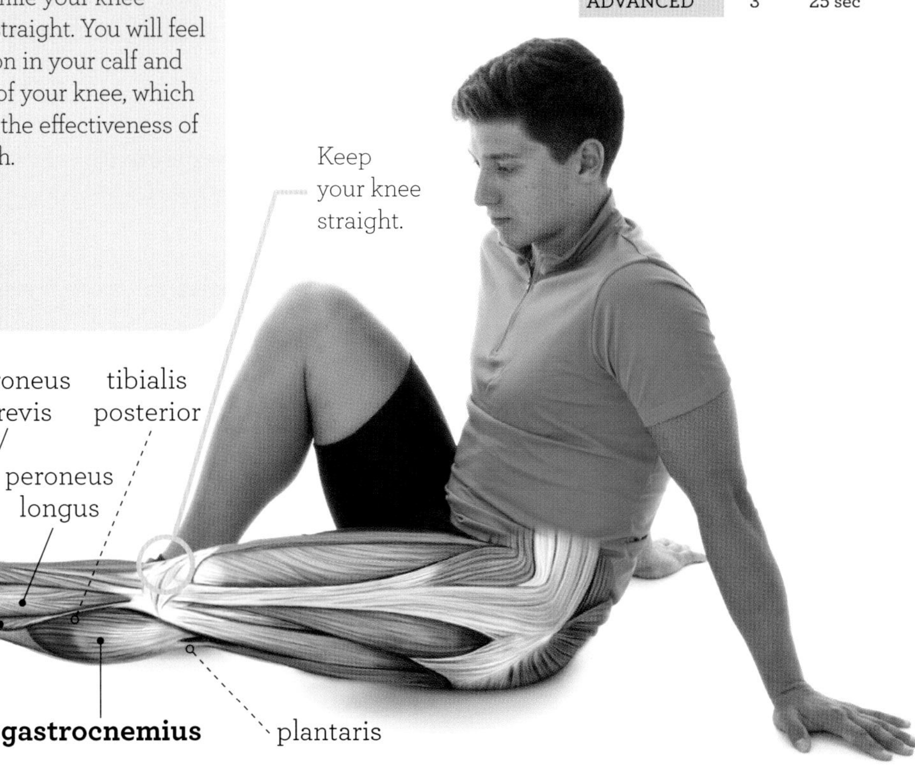

CAUTION

Even though gastrocnemius stretches involve no great risk, because of the strength of the muscle, it's a good idea to maintain continuous communication between the two parties in order to get the greatest benefit form the exercise and minimize possible undesirable effects.

INDICATION

For cyclists who experience strain or muscle tension in the calf or discomfort or nonacute tendonitis in the Achilles tendon. Also for all cyclists, because of the crucial role of the muscles responsible for plantar ankle flexion in pushing the pedals.

LEVEL	REPS	DURATION
BEGINNER	2	20 sec
INTERMEDIATE	2	25 sec
ADVANCED	3	25 sec

Unilateral Pull with Towel

Keep your knee straight and the back of your thigh in contact with the floor.

START
Before starting the exercise, take a small towel or similar item you can use for pulling on your foot. Sit down with one leg straight and the other one drawn up, so that the front of your thigh touches your chest, your knee is bent, and the sole of your foot touches the floor. Pass the towel around the sole of the forward foot, even with the metatarsophalangeal joints, and hold one end in each hand. Your ankle should remain in a neutral position or in plantar flexion.

TECHNIQUE
Pull on the towel simultaneously with both hands. This will produce dorsal flexion of the ankle and the resulting stretch of the sural triceps. Keep your ankle straight as you pull on the towel to affect the gastrocnemius in particular. You will feel the muscle tension in the calf and the back of your knee.

CAUTION

Try to keep your upper body perpendicular to the floor and try not to lean it during the pull. This way, the gastrocnemius will yield to the stretch without producing tension in the lumbar vertebrae.

INDICATION

For cyclists who feel discomfort in the Achilles tendon or experience nonacute tendonitis, as well as for those who exhibit muscle tension in the calf. Also for cyclists in general, because of the intense work performed by the gastrocnemius in pedaling, especially on long rides or in response to climbs.

Lunge with Dorsal Ankle Flex

START

Stand with your upper body perpendicular to the floor, your hands on your waist, and one foot slightly ahead of the other. Keep your knees straight and look straight ahead. Be sure you have a clear space in front of you to take a few strides or lunge forward.

TECHNIQUE

Lunge forward with the front foot, while keeping the rear foot in its original position. The latter should make full contact with the sole, including the heel, to produce the proper stretch. Bend the forward knee slowly while keeping the other one straight. As your center of gravity moves downward, the rear ankle will increase its dorsal flexion.

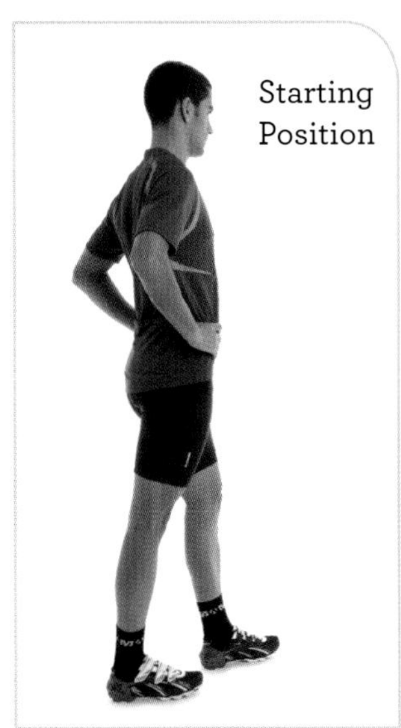
Starting Position

LEVEL	REPS	DURATION
BEGINNER	2	20 sec
INTERMEDIATE	2	25 sec
ADVANCED	3	25 sec

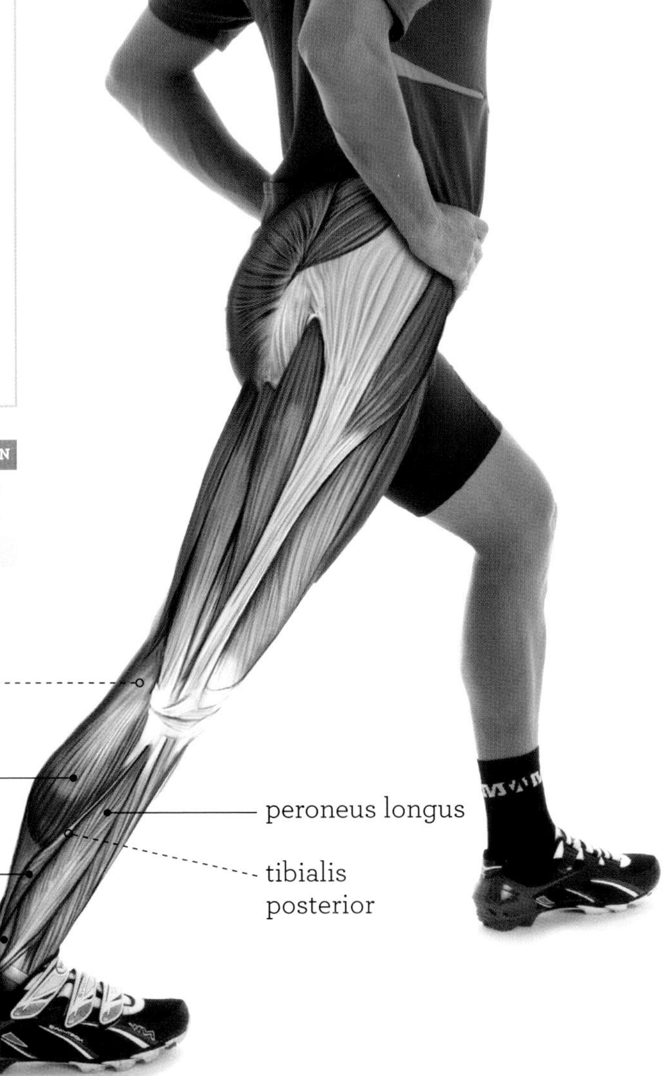
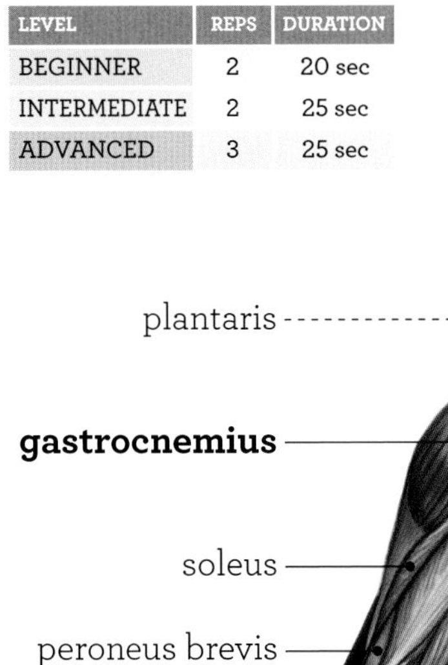

Keep your heel on the floor.

CAUTION

Start from a stable position, so that you perform the exercise safely and without interruption. Keep the rear knee straight to emphasize the stretch in the gastrocnemius.

INDICATION

For cyclists who commonly do long rides, who feel tension or muscle strain in the calf or discomfort in the Achilles tendon, as well as for cyclists in general, because of the involvement of the muscles responsible for plantar flexion of the ankle in pedaling.

Dorsal Ankle Flex on Pedal

LEVEL	REPS	DURATION
BEGINNER	2	20 sec
INTERMEDIATE	2	25 sec
ADVANCED	2	30 sec

Lower your heel as far as possible while keeping your foot on the pedal.

START

Stand straddling your bicycle, hold the handlebars with both hands, and squeeze your brake levers. Place one pedal in its lowest position and rest the front of your foot on it, clipped in if possible. At this time there will be a slight bend in your knee and your ankle should be in a neutral position.

TECHNIQUE

Straighten the knee of the leg being stretched and lower your heel while keeping your foot on the pedal. This will require you to force the dorsal flex of your ankle, and thus the stretch in the muscles that perform plantar flexion. You will feel the tension in your calf, and it will increase as you lower your heel with respect to your toes.

CAUTION

Because of the pressure you will exert on the pedal, it will have to be at the bottom of the stroke, and your contact will have to be firm and sure to keep your foot form slipping off and getting hit.

INDICATION

For cyclists who experience tension or muscle strain in the calf or discomfort in the Achilles tendon. Also for cyclists in general, because of the force exerted by the gastrocnemius in pedaling.

Unilateral on Step

START
Stand on a step, a curb, or some other raised surface. Move one foot back so that it touches the edge of the step at the metatarsophalangeal joint. The middle part of your foot and your heel will hang over the edge. The knee of the rear leg should be straight, and the other leg can be bent slightly.

TECHNIQUE
Bend the knee of the forward leg so that your center of gravity moves downward and your rear ankle remains in dorsal flexion. For this to happen, your heel will need to move downward as you maintain contact with the front of your foot. The knee of the leg being stretched should remain straight so that the stretch affects the gastrocnemius more. You will feel the tension in your calf and the back of your knee and ankle.

Starting Position

LEVEL	REPS	DURATION
BEGINNER	2	20 sec
INTERMEDIATE	2	25 sec
ADVANCED	3	25 sec

plantaris
gastrocnemius
soleus
tibialis posterior
peroneus brevis
peroneus longus

Keep your knee straight.

CAUTION
Do this exercise on a slight rise, such as a step or a curb, and make sure that the foot being stretched touches it securely.

INDICATION
For cyclists who experience strain or muscle tension in the calf, who suffer from nonacute tendonitis in the Achilles tendon, or who regularly ride long distances. Also for cyclists in general, because of the involvement of the gastrocnemius in pedaling.

Shallow Squat with Hands on Hips

LEVEL	REPS	DURATION
BEGINNER	2	20 sec
INTERMEDIATE	2	25 sec
ADVANCED	2	30 sec

START

Stand with one foot forward, your upper body perpendicular to the floor, and your knees straight. Your feet should be parallel and point straight ahead. You can place your hands on your hips for improved comfort in the execution phase.

TECHNIQUE

Bend both knees to lower your center of gravity. Keep your upper body perpendicular to the floor and your feet in the same place. They should remain aligned and with the heels on the floor. As your center of gravity moves downward, the stretch in the soleus and other muscles will increase, and you will feel the muscle tension in your calf. Even though you will feel a certain amount of stretch in both lower legs, it will be more pronounced in the rear one.

CAUTION

This stretch involves no particular risk; but, when you plan your stretching session, you should keep in mind that when you stretch the gastrocnemius, you are also stretching the soleus.

INDICATION

For all cyclists, because of the involvement of the soleus in pedaling, particularly in moving the pedals forward and downward. Especially recommended for people who experience or have experienced nonacute tendonitis of the Achilles tendon or muscle tension in the calf.

Seated Foot Pull

Starting Position

START
Sit down, bend the knee of the leg being stretched, and place your heel on the floor. Lean your upper body forward, hold the sole of your foot with both hands, and keep your ankle in a neutral position. The other leg can be stretched out next to the first one or held more to the side, but it should not interfere with the arm position.

TECHNIQUE
Pull back on your toes while keeping your heel in contact with the floor, forcing the dorsal flexion of the ankle. As this becomes more pronounced, you will feel the tension in your calf and the rear of your ankle, but this feeling will not be as pronounced as in other stretches. This is no cause for concern, because the tension varies based on the muscle group and the exercise being performed.

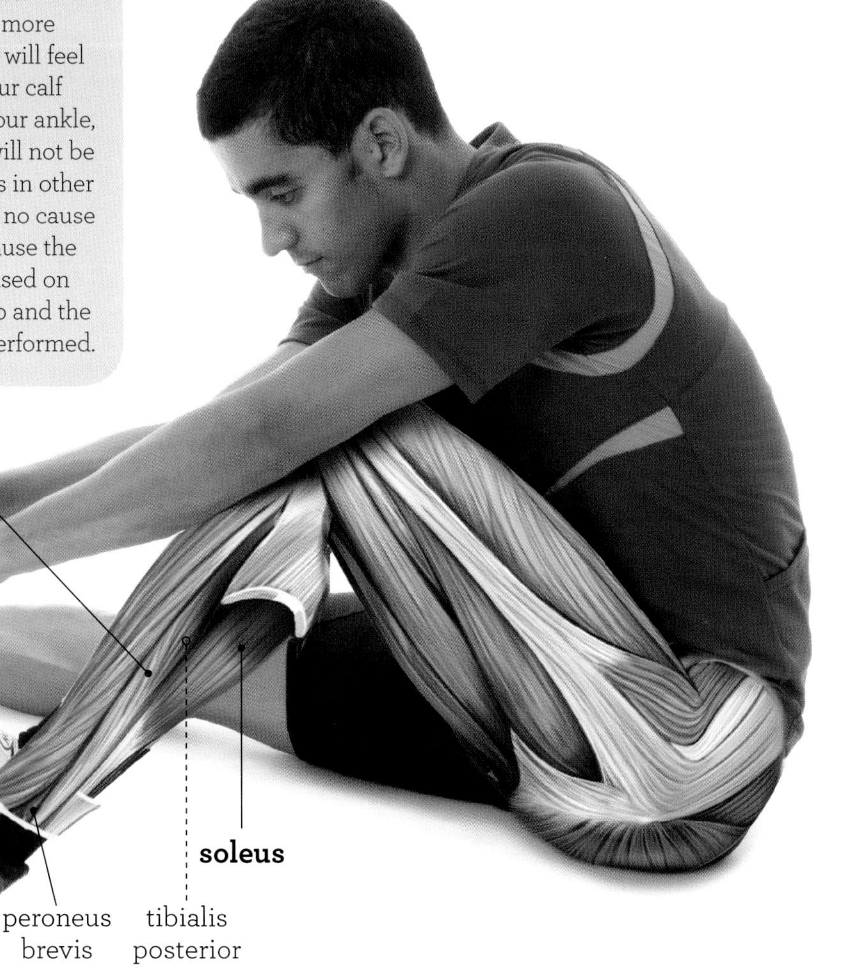

Keep your heel on the floor without sliding it toward your hip.

LEVEL	REPS	DURATION
BEGINNER	2	20 sec
INTERMEDIATE	2	25 sec
ADVANCED	2	30 sec

CAUTION

Try not to move your heel from its original position, becaue moving your foot toward your hip may give the false sensation of greater movement, but, in fact, it does not contribute to the proper performance of the exercise.

INDICATION

For cyclists who experience tension in the calves or the Achilles tendon, who mainly do long rides, or who ride very hilly routes. Also for cyclists who ride regularly, because of the involvement of the muscles used in plantar flexion of the ankle in pedaling.

Unilateral Deep Squat

Starting Position

START
Squat on one foot and one knee so that the front of one leg touches the floor. Lean your upper body rearward so you can place your hands on the floor and practically sit on your rear foot.

TECHNIQUE
Lean your upper body forward and slide your hands in the same direction along the floor, trying to reach as far as possible. Your support foot should remain in its original location, and the advance of the rest of your body will force your ankle into maximum dorsal flexion. Go as far as you can without lifting your heel from the floor, because this would reduce the effectiveness of the exercise.

soleus
tibialis posterior
peroneus longus
peroneus brevis

Keep your heel on the floor.

LEVEL	REPS	DURATION
BEGINNER	2	20 sec
INTERMEDIATE	2	25 sec
ADVANCED	2	30 sec

CAUTION

If you feel discomfort in your ankle at any time, especially in the front, reduce the intensity of the stretch or stop doing it. Remember that in this exercise the muscle tension will not be as evident, and continuing the movement indefinitely may only contribute to hurting your ankle.

INDICATION

For cyclists in general, because of the use of the soleus in pushing the pedals forward and down, especially on long rides or going up significant hills. Also for cyclists who are prone to experiencing muscle or tendon discomfort in the calf and the back of the ankle.

Foot Pull with Knee Bend

START

Assume the knight's position—in other words, kneel on one foot and one knee. The hip and knee of the support foot will be bent around 90°, and the other thigh will be in line with your upper body. Lean your upper body toward the forward foot and hold your toes with both hands at the metatarsophalangeal joints.

TECHNIQUE

Pull on your toes with both hands to produce the maximum possible dorsal flexion in your ankle. Your heel should remain in contact with the floor and in its original position. Do the movement slowly, but apply considerable force. In this exercise, the tension will not be felt as clearly as in others, but you will feel it in your calf if you do it correctly.

LEVEL	REPS	DURATION
BEGINNER	2	20 sec
INTERMEDIATE	2	25 sec
ADVANCED	2	30 sec

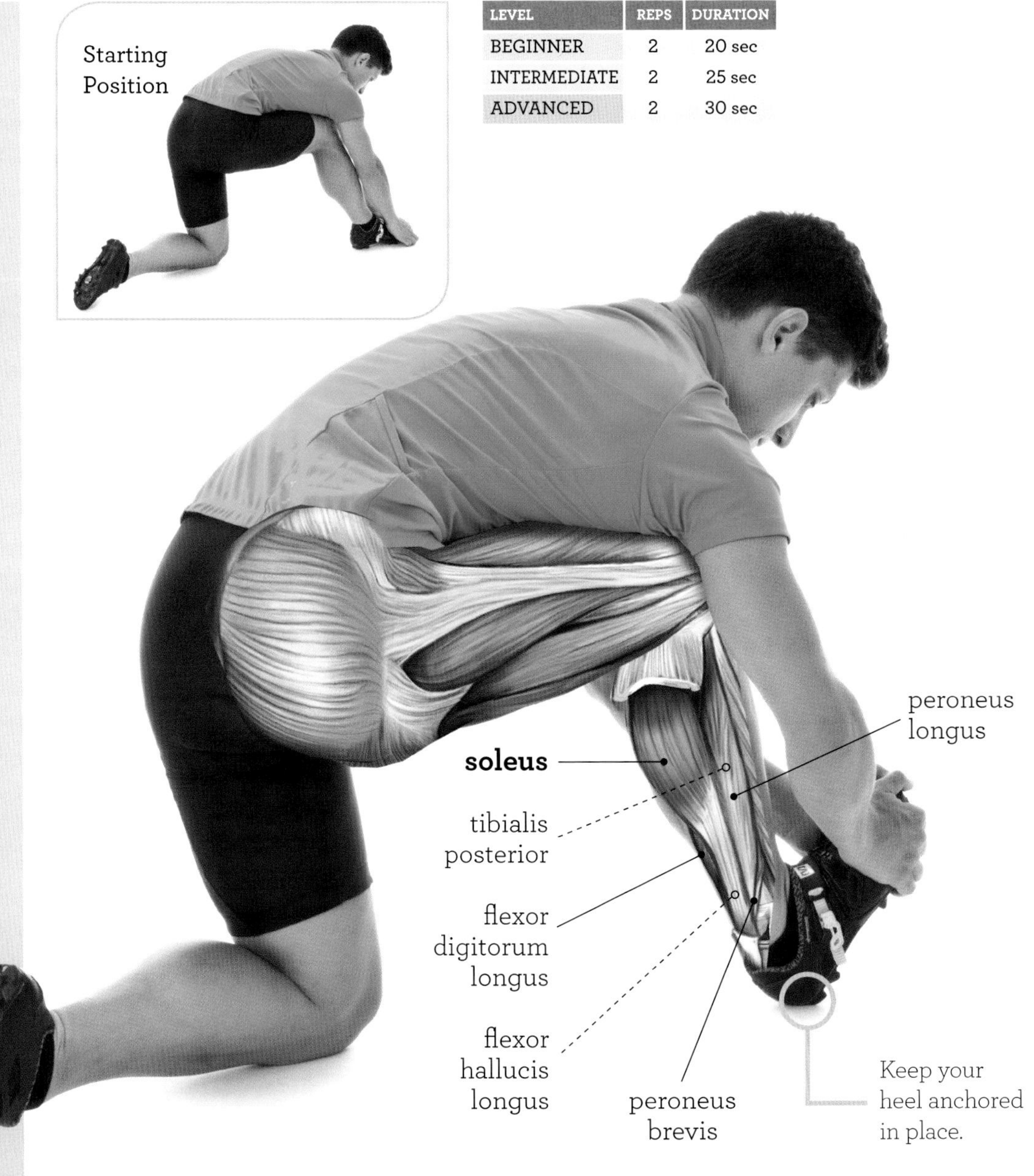

CAUTION

Use a mat or something similar to avoid resting your knee directly on the floor. Remember that you should stop if you feel any joint pain in your ankle that goes beyond the appropriate feeling for a stretch.

INDICATION

For cyclists in general, because of the involvement of the muscles responsible for plantar flexion in pushing the pedals forward and down while pedaling. Especially for those who experience strain or muscle tension in the calf. Also for cyclists who are prone to discomfort in the Achilles tendon, who do very long rides, or who climb steep, long hills.

Toe Pull

LEVEL	REPS	DURATION
BEGINNER	2	15 sec
INTERMEDIATE	2	20 sec
ADVANCED	2	25 sec

START

Sit down with one leg straight and the other one crossed over it. To improve the effectiveness of the stretch, remove your shoe from the top leg. Hold your heel with one hand and your toes with the other one on top of the metatarsophalangeal joints, so that you produce a proper and broad utilization of this joint as you perform the exercise. Your ankle should be in a neutral position or in slight dorsal flexion.

TECHNIQUE

Pull rearward on your toes while you hold your heel stationary and keep your ankle in place on your thigh. Try to reach the maximum extension in the metatarsophalangeal and intertarsophalangeal joints. As this extension occurs, you will feel tension in your joints and the sole of your foot, and you will feel the tension in the relevant foot by touch.

CAUTION

Stress the extension of the metatarsophalangeal joints more than the intertarsophalangeal joints, because of the relative fragility of the latter, and pay attention to any sensation of pain in your toes and joints, which should be a signal to reduce the intensity of the stretch or to stop the exercise.

INDICATION

For cyclists who experience tension in the soles of the feet, which may result from the work involved in pushing the pedal with the front of the foot.

syncros

BICYCLE STRETCHES

THE BASICS OF STRETCHING ON THE BICYCLE

Sometimes, cycling events go on for hours, so riders can begin to feel discomfort or muscle strain on the bike and have no time or opportunity to stop, massage, and stretch the affected area(s). In these situations, the discomfort can increase in intensity, and even keep the rider from finishing the event or from doing so in good shape.

The most common discomfort is the muscle strain that arises with fatigue, which may cause pain and functional disability. In instances where getting off the bicycle is not an option, there are some stretches that can help to alleviate muscle strain and discomfort without having to take a break (with the time loss that this would involve in a competition).

In other sports in which there are breaks or rest periods, we occasionally see athletes receive help from massage therapists. In tennis matches, we sometimes see players receive massages or do stretches during the breaks between sets. In soccer games, we sometimes see players stretch during halftime in order to encourage recovery so that they can handle more successfully the subsequent half of the game.

These massages and stretches, like the ones that are done after exercising, produce a basic recuperative effect in muscle relaxation and improved venous return, which makes it possible to eliminate residual substances from the tissues, such as lactic acid. This, in turn, reduces muscle congestion and allows the arrival of oxygenated blood loaded with nutrients at the muscle tissue, which allows for added activity

Muscle congestion due to intense exercise during a long climb can limit venous return and proper recovery of the blood that accumulates in the muscle.

time and improves performance. A combination of massage or self-massage and stretching during long athletic events can contribute significantly to setting personal records.

From this point of view, stretches on the bicycle can help in situations where there is no other choice, and they offer a possibility of pedaling the extra distance needed to finish the event, but they are not the safest option. Their drawback is that, to a greater or lesser degree, they detract from stability or visibility in riding, so all cyclists need to assess whether they have the ability and the need to do them. A person who uses the bike sporadically or purely for recreational purposes and personal satisfaction will get better results, especially greater safety, by doing conventional stretches, and their best option will be to get off the bike and do the exercises presented in this book, with the repetitions and times indicated, because they are always more effective than the stretches that are done while riding.

Only cyclists with very good control of the bike and special needs connected to competition should use the stretches on the bike, because they sometimes involve changing the grip, removing the hands from the bars, or looking away form the road for a few seconds, which of course is not within everyone's capability. In most cases, this involves running an unnecessary risk. For this reason, the indicated times have been adapted for the sake of safety above effectiveness, and they have been reduced considerably, practically turning some of the exercises into dynamic stretches, which probably should be done in higher numbers than indicated or with greater frequency in order to improve results.

In addition, even more experienced cyclists who can do these exercises safely should do them only on flat terrain and under optimal road and competition conditions, so these exercises do not apply to off-road trails, rough surfaces, steep climbs or downhills, and other situations involving adverse riding conditions.

It is appropriate to remember that not even the best results in competition can make up for a fall and the harmful consequences that may occur. So remember that the following exercises should be done only if you are an experienced cyclist and can do them under safe conditions.

Exercises done on the bicycle involve a significant risk of falling, and all cyclists must assess whether they need these exercises and whether they are capable of doing them safely.

Self-assisted Neck Bend on the Bicycle

START
Straighten your elbows and assume the most erect position possible without letting go of the handlebars and while maintaining contact with the saddle. Place one hand on your head, which you should keep erect so that you have control over your direction of travel.

TECHNIQUE
Pull your head to the side, attempting to touch your ear to your shoulder through lateral neck flexion. You will feel muscle tension in the side of your neck opposite the pulling hand, which will indicate that the stretch is being done correctly. Keep the bike moving through momentum or gentle pedaling while you do this exercise.

LEVEL	REPS	DURATION
BEGINNER	1	10 sec
INTERMEDIATE	2	10 sec
ADVANCED	3	10 sec

CAUTION

Watch where you are going and remember to adapt the times to the cycling conditions, which should always determine the duration of the exercise. The repetitions and times indicated here are merely guidelines.

INDICATION

For cyclists who experience tension in the neck due to their position on the bicycle, which especially affects road and track racers, because of their more aerodynamic position.

Neck Bend on the Bicycle

START
Sit up while touching the saddle and holding the handlebar with one or both hands. Be sure to do this at low speed and when the stretch ahead of you is straight, clear, and smooth.

TECHNIQUE
Bend your neck as far as you can for a brief instant to stretch the muscles in the back of your neck. You will feel tension in the area of your cervical and upper thoracic vertebrae. You can use one hand to increase the intensity of the stretch. This movement can relax neck strain momentarily and allow you to continue riding.

LEVEL	REPS	DURATION
BEGINNER	1	1–2 sec
INTERMEDIATE	2	1–2 sec
ADVANCED	3	1–2 sec

CAUTION

In this exercise, you have to take your eyes off the road for a few seconds, so keep it very short and make sure that the road ahead of you is clear. It is better to do this exercise more frequently and limit it to a brief instant each time.

INDICATION

For cyclists who experience tension in the neck extensor muscles, because of their position on the bicycle, especially if the event in question is long or done in stages, as with road races.

Posterior with Arm Crossed in Front

START
Let go of the handlebars and sit up straight on the bike. Cross one forearm in front of your abdomen and hold the elbow with your other hand so that both elbows are bent around 90°. Make sure that the road ahead of you is clear, because you are riding with no hands on the handlebars.

TECHNIQUE
Pull on your elbow so that your arms remain crossed in front of your chest in maximum frontal adduction. The rear part of the deltoid will be stretched, with an attendant sensation of muscle tension. Hold the position for a few seconds, and always give priority to the need to keep control of your bike.

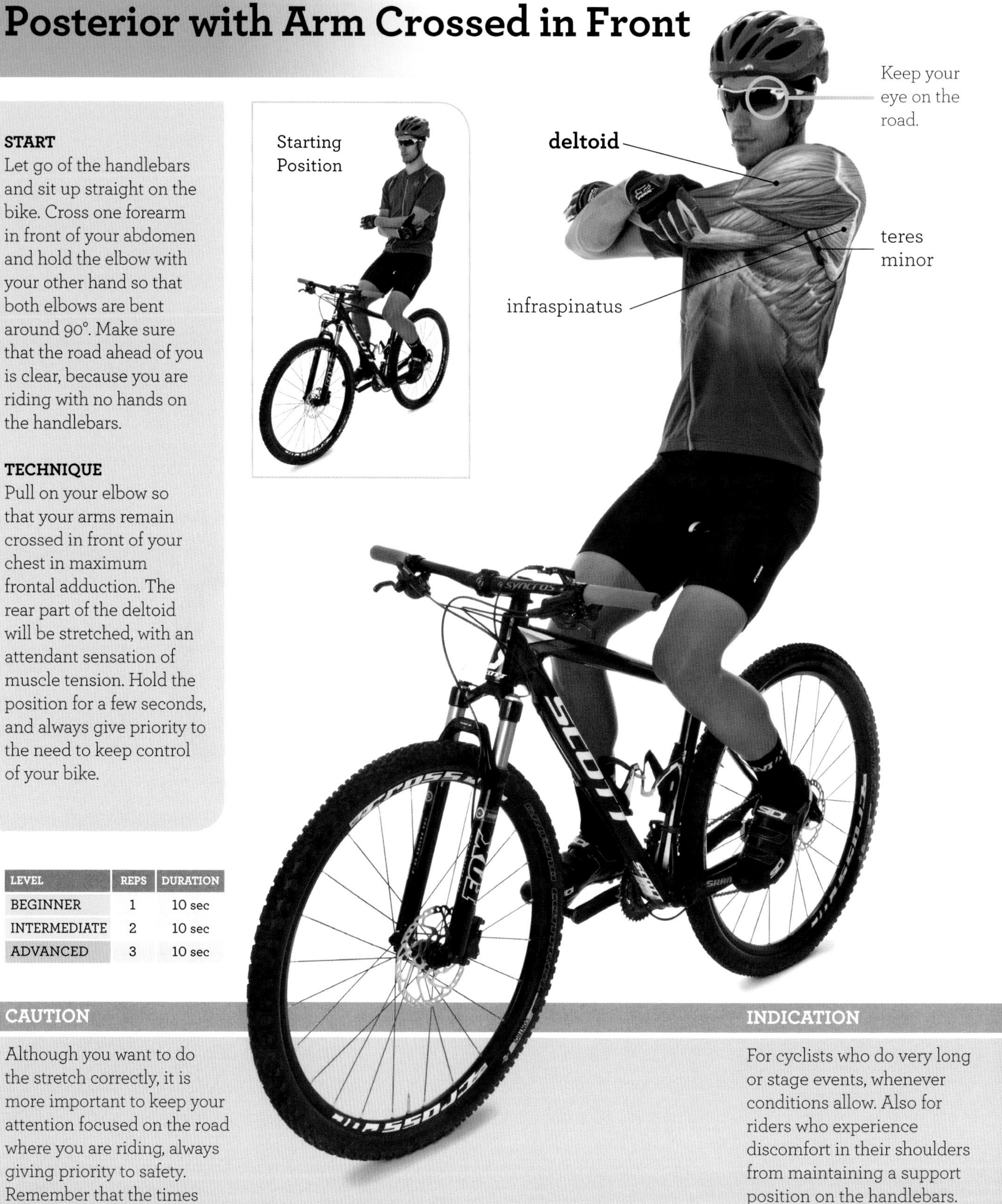

LEVEL	REPS	DURATION
BEGINNER	1	10 sec
INTERMEDIATE	2	10 sec
ADVANCED	3	10 sec

CAUTION

Although you want to do the stretch correctly, it is more important to keep your attention focused on the road where you are riding, always giving priority to safety. Remember that the times specified are only rough guidelines.

INDICATION

For cyclists who do very long or stage events, whenever conditions allow. Also for riders who experience discomfort in their shoulders from maintaining a support position on the handlebars.

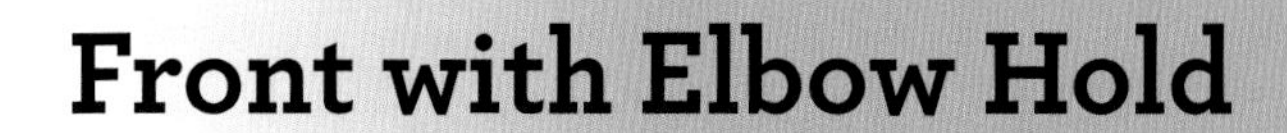

Front with Elbow Hold

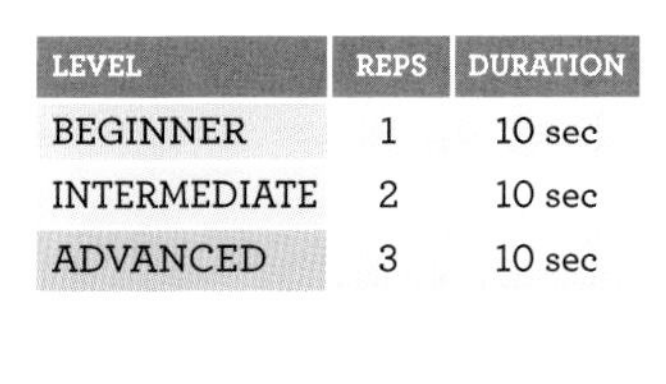

LEVEL	REPS	DURATION
BEGINNER	1	10 sec
INTERMEDIATE	2	10 sec
ADVANCED	3	10 sec

START

While riding with both hands on the handlebars, make sure the road is clear in front of you for a fair distance, preferably on a straightaway without any steep hills. This will allow you to keep pedaling gently and at a relatively low speed, which will help balance the bike once you let go of the handlebars.

TECHNIQUE

Let go of the handlebars and keep your upper body perpendicular to the ground. Place your forearms behind your upper body and try to grasp each elbow with the opposite hand. If you can't reach your elbows, you can hold your forearms instead. Pull simultaneously with both hands, forcing shoulder adduction. You will feel the tension in your shoulders.

Starting Position

CAUTION

Don't take your eyes off the road at any time, and adapt the times to the requirements of controlling the bike. Remember that the times indicated are just rough guidelines, and, if you need to interrupt a stretch before the recommendation in the chart, you can always do another repetition when the circumstances allow.

INDICATION

For cyclists who experience discomfort in the shoulders, especially during long or stage events, provided that they have excellent control of the bike and the road conditions allow.

Front with Arms Behind on Bicycle

START
Make sure that the road ahead of you is clear, and let go of the handlebars. Sit up so that your upper body is perpendicular to the ground and both arms hang down with your elbows straight. Keep your eye on the road ahead of you so that you can react to any danger.

TECHNIQUE
Clasp your hands behind your back, straighten your elbows, and simultaneously perform retropulsion with your shoulders until you feel muscle tension in your chest and the front of your shoulders. Hold the position for a few seconds, provided that the road conditions allow you to do so.

LEVEL	REPS	DURATION
BEGINNER	1	10 sec
INTERMEDIATE	2	10 sec
ADVANCED	3	10 sec

CAUTION

Remember to give priority to safe riding over performing the exercise, and don't take your eyes off the road. Try to do the stretch on a flat section or when pedaling does not require too much effort.

INDICATION

For cyclists who experience discomfort in the shoulders while participating in riding events, especially if they are very long or they involve several stages.

Wrist Extension on Handlebars

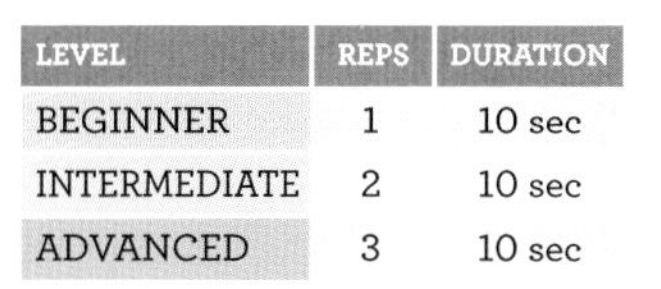

LEVEL	REPS	DURATION
BEGINNER	1	10 sec
INTERMEDIATE	2	10 sec
ADVANCED	3	10 sec

Starting Position

flexor carpi ulnaris

flexor digitorum superficialis

flexor digitorum profundus

palmaris longus

flexor carpi radialis

Extend both wrists and hold the handlebars with your fingers.

START
Take a standard grip on the top of the handlebars, keeping your wrists in a neutral position, your elbows straight but not locked, and your shoulders in antepulsion or flexion. Maintain your contact point on the saddle and look straight ahead to make sure that there is no rough pavement in the section ahead of you.

TECHNIQUE
Rotate your wrists, placing them in extension as if you were operating the throttle of a motorcycle, so that your wrists move downward and you are touching the handlebars only with your fingers. The flexor muscles of the wrist will stretch, and you will feel tension in them and in the front part of your forearms.

CAUTION

Avoid holding the handlebars with your fingertips, because the main support should be on the proximal phalanges. This will assure an effective stretch and a sufficiently firm grip on the handlebar so that you won't lose your grip while doing the exercise if you are riding on a smooth surface.

INDICATION

For cyclists who experience discomfort in their wrists, hands, and forearms while participating in cycling events, especially if they are long or run in stages.

Finger Extension on the Bicycle

START
In this exercise, you will have to let go of the handlebars with both hands; therefore, before you begin, make sure that the road ahead of you is level, as well as straight and smooth. Try to use a cadence that doesn't require too much pedaling force.

TECHNIQUE
Let go of the handlebars so that your upper body is perpendicular to the ground or slightly leaning to the rear. Place your hands in front of you by flexing your shoulders, and fold your hands with your palms facing forward. Straighten your elbows to increase the finger and wrist extension. You will feel the effect of the stretch mainly in the bases of your fingers.

LEVEL	REPS	DURATION
BEGINNER	1	10 sec
INTERMEDIATE	2	10 sec
ADVANCED	3	10 sec

CAUTION

Keep your eyes on the road and stop the stretch to hold the handlebars whenever circumstances require. Always prioritize your safety, and remember that you can do the stretch later on if the conditions allow.

INDICATION

For all cyclists, because of the work that they do with their hands in holding the handlebars and using the brake levers and gear shifters while cycling. Especially for those who feel discomfort or numbness in their hands and wrists.

Wrist and Finger Flex on the Bicycle

LEVEL	REPS	DURATION
BEGINNER	1	10 sec
INTERMEDIATE	2	10 sec
ADVANCED	3	10 sec

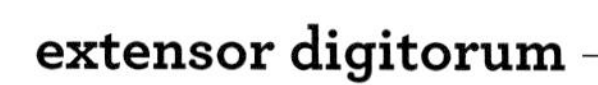

START

Let go of the handlebars with both hands and assume an upright position on the bicycle, with your upper body perpendicular to the ground. Bend your elbows to 90° and cross your forearms and your hands in front of your abdomen. Hold one hand with the other, as if you were trying to cover the first one, which should remain with your fingers bent and the fist clenched. Keep your eyes on the road to make sure that the section ahead of you is free of obstacles.

TECHNIQUE

Use the gripping hand to force the flexion of the other wrist, while keeping the hand being stretched closed in a fist. Do the movement slowly, and you will feel the tension in the back of the wrist and hand. Remember that you can do this exercise as many times as you wish, even after long breaks between sets.

CAUTION

As with all exercises that require letting go of the handlebars, don't do this if it is not safe or if it is not necessary, and, if you do it, make sure that the section of road ahead of you is clear.

INDICATION

For cyclists who feel discomfort in their wrists or pain or numbness in their hands and fingers during a competition. In most cases, these conditions are due to the long-term support on the handlebars and the compression to which the median and ulnar nerves are subjected.

Spine Extension I

LEVEL	REPS	DURATION
BEGINNER	1	10 sec
INTERMEDIATE	2	10 sec
ADVANCED	3	10 sec

START
Stand up from the saddle so that your body weight is supported on the pedals and the handlebars, as if you were going to pedal while standing. Keep your elbows and your back straight, and the crank arms parallel to the seat tube. Your upper body should lean forward slightly, and the leg on the lower pedal should be nearly perpendicular to the ground, with your knee straight.

TECHNIQUE
Move your hips forward toward the handlebar stem, while keeping your elbows straight and your grip on the handlebars, so that your shoulders are nearly in the same position as at the start. Try to straighten your lumbar vertebrae to compensate for the nearly constant bend in the position that's used on the bicycle.

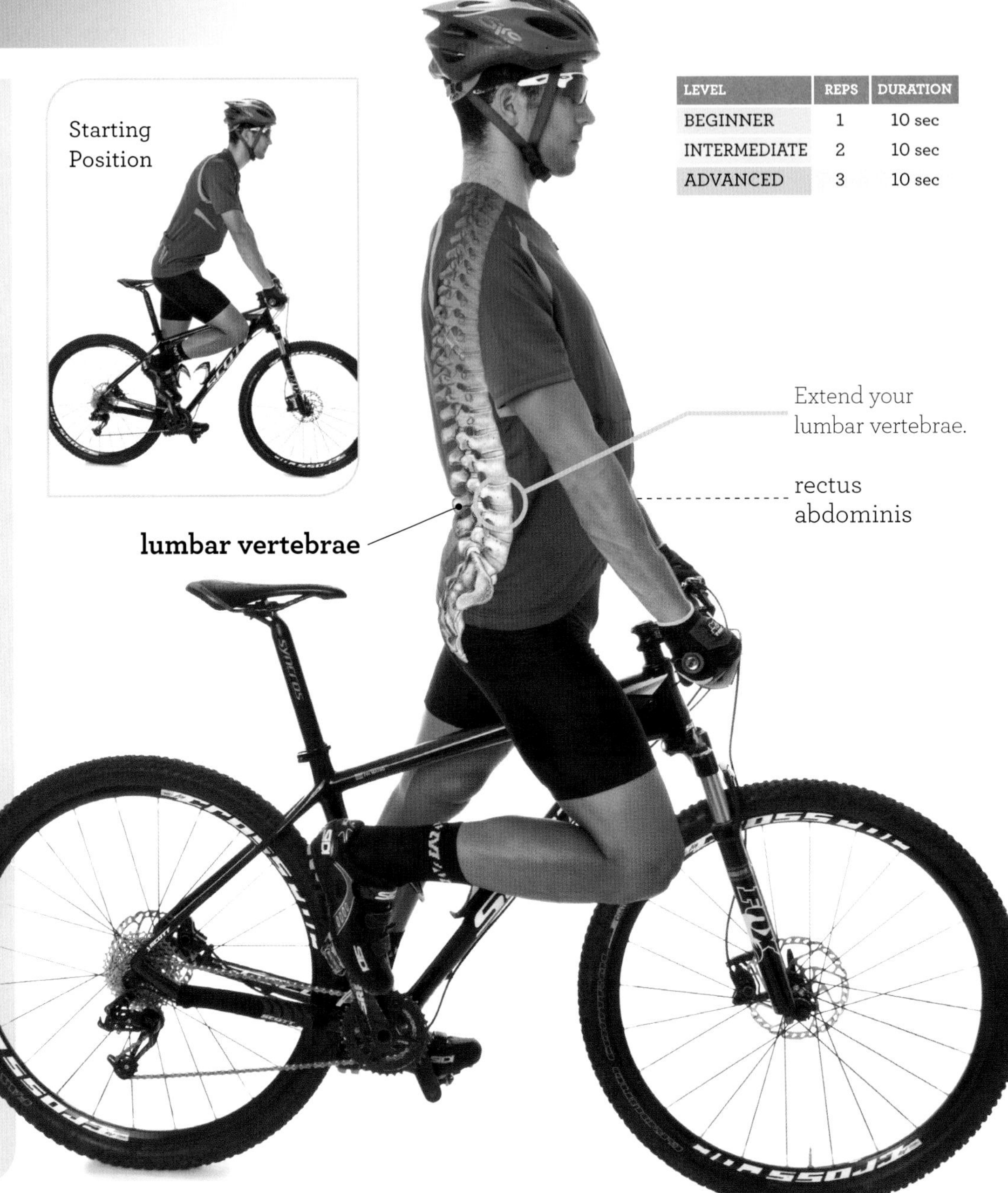

CAUTION

In this exercise, even though you keep both hands on the handlebars, your eyes on the road, and your feet on the pedals, the position gradually becomes more unstable and less comfortable, because of the weight shift forward, so you should try to do this exercise only when the section of road where you are riding offers the best conditions.

INDICATION

Especially for cyclists who experience discomfort in the lumbar vertebrae during long events. This exercise is not really a stretch, strictly speaking, because it doesn't aim to stretch any muscle group, but rather seeks to reduce pressure on the spine, relax the entire lumbar spinal column, stretch the intervertebral disks, and alleviate the tension produced in adjacent areas of the body.

Spine Extension II

START
Hold yourself on the bicycle with both hands on the handlebars and make sure you have a safe, flat section of road without any obstacles ahead of you while performing this exercise. It is best to do this exercise on a section of road where there is no need to exert great force on the pedals, because you have to let go of the handlebars to perform the exercise.

TECHNIQUE
Let go of the handlebars and sit up on the saddle, with your upper body perpendicular to the ground. Place your hands over your kidneys and extend your spine as far as possible, expecially in the lumbar region, keeping your upper body perpendicular to the ground. This will require a certain amount of anteversion of the pelvis.

LEVEL	REPS	DURATION
BEGINNER	1	10 sec
INTERMEDIATE	2	10 sec
ADVANCED	3	10 sec

CAUTION

As in all cases where you let go of the handlebars, you will have to adjust the performance and the duration of this exercise to the riding conditions and the road. Consequently, you can shorten the times indicated and do the stretches more frequently.

INDICATION

This exercise, like the previous one, does not aim to stretch a specific muscle group, but rather to alleviate the pressure on the spine, particularly the lumbar region and related areas, which results from the pedaling position, especially if it is very aerodynamic and the event is long.

Knee Pull to Chest on the Bicycle

START

Make sure you have a flat stretch of road ahead of you, or one with a slight downhill, on which you can stop pedaling for a few seconds without losing momentum on the bike. Place the crank arms in line with the seat tube, let go of the handlebars with one hand, and place the hand onto the front of the knee on the same side, which should be the one that's on the top pedal and bent the most.

TECHNIQUE

Pull on your knee so that your foot comes off the pedal and your hip flexes as much as possible. The front of your thigh should come close to your chest for the stretch to be effective. Remember that the muscle tension will be much lower than what you can feel in conventional stretches, but it is still sufficient, given the situation in which this stretch is done.

LEVEL	REPS	DURATION
BEGINNER	1	10 sec
INTERMEDIATE	2	10 sec
ADVANCED	2	10 sec

Starting Position

CAUTION

This exercise changes your balance on the bike perceptibly by moving the support points to one side, so you should do this exercise slowly and only if your are certain you can control the bike at every instant.

INDICATION

For cyclists who feel muscle strain in the gluteus muscles in competition. This is most common in time racing and climbing events, especially long ones.

Knee Bend with Rear Hold

START

Make sure that the section of road ahead of you is level or slightly downhill so you can maintain your speed, even if you are not pedaling for a few seconds, because you will have to free up one foot and one hand to do this exercise.

TECHNIQUE

Take one foot off the pedal and hold it by the instep using the hand on the same side. Pull it rearward by bending your knee and extending your hip as far as possible at the same time. This will be sufficient to produce a moderate stretch in the quadriceps femoris, which you will feel slightly in the front part of your thigh.

Starting Position

LEVEL	REPS	DURATION
BEGINNER	1	10 sec
INTERMEDIATE	2	10 sec
ADVANCED	2	10 sec

CAUTION

This exercise changes your balance on the bicycle significantly, creating a potentially dangerous situation; therefore, you should do the exercise only until you feel a slight stretch and no farther. Remember that you are merely trying to slightly alleviate muscle tension.

INDICATION

For competitive cyclists who experience muscle tension in the front of the thigh, which commonly occurs in long climbs and on long rides, as in time-racing events and most track events.

Dorsal Ankle Flexion on Pedal

START
Stand up out of the saddle and place the crank arms in line with the seat tube. You will have to move your center of gravity forward, closer to the handlebar stem, and support your body weight on the pedals and the handlebars. Make sure you have a clear, smooth section of road ahead of you.

TECHNIQUE
Shift a good portion of your weight onto the lower pedal and perform dorsal flexion with your ankle, while keeping your knee straight. Your body weight will accentuate this movement. This exercise should be done only if your feet are clipped in to the pedals, to avoid the risk of your foot sliding off the pedal.

LEVEL	REPS	DURATION
BEGINNER	1	10 sec
INTERMEDIATE	2	10 sec
ADVANCED	3	10 sec

CAUTION

Do this exercise when your route of travel is level or slightly downhill, and when the road is smooth, because you will have to stop pedaling for a few seconds, and the weight distribution on the bicycle will change.

INDICATION

For cyclists who feel tension or discomfort in the muscles responsible for plantar flexion of the ankle while participating in an event. Climbers and long-distance cyclists are most susceptible.

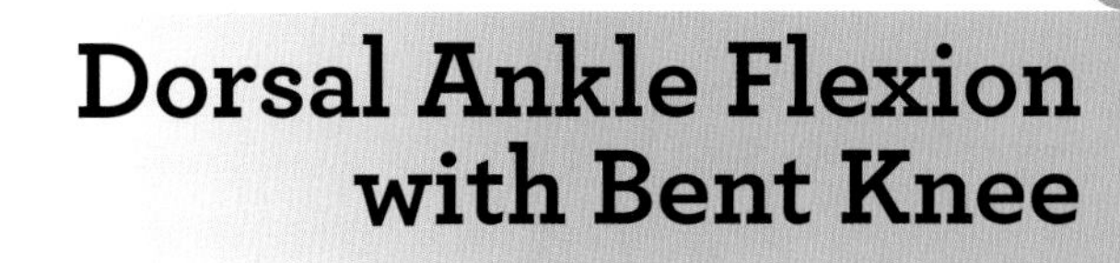

Dorsal Ankle Flexion with Bent Knee

Starting Position

Simultaneously perform the dorsal ankle flexion and the knee bend.

soleus

peroneus longus

tibialis posterior

peroneus brevis

START

Stand up out of the saddle and move your center of gravity forward, closer to the handlebars. The crank arms should be in line with the seat tube, and the ankle of the lower foot should be in a neutral position. Make sure the road ahead of you is level or slightly downhill to continue riding safely, even though you stop pedaling for a few seconds.

TECHNIQUE

Bend the knee of the leg being stretched and place the ankle in dorsal flexion. Shift a good part of your weight toward this side to accentuate the dorsal ankle flexion, and hold the position for a few seconds. You will feel the muscle tension in your calf.

LEVEL	REPS	DURATION
BEGINNER	1	10 sec
INTERMEDIATE	2	10 sec
ADVANCED	3	10 sec

CAUTION

Use this technique only if you use clip-in pedals, because your foot must remain firmly anchored to the pedal while doing the stretch.

INDICATION

For cyclists who feel muscle discomfort in the calf during competition, especially if they do steep, long climbs or if the event is very long.

Alphabetical Index of Muscles

Bibliography

Cavill, Nick and Adrian Davis, *Cycling & Health: What's the Evidence?*, Longon Cycling, 2007.

Hutson, Warren G., "Cycling Injuries—Prevention and Treatment," *SportEX Dynamics*, 2006, Issue 10, p. 19.

Impellizzeri, Franco M. and Samuelle M. Marcora, "The Philosophy of Mountain Biking," *Sports Med.*, 2007, Vol. 37, No. 1, pp. 59–71.

Moore, Fran, "Correct Bicycle Set-up to Minimize the Risk of Injury." *SportEX Dynamics*, 2008, Issue 37, pp. 6–9.

Schultz, Samantha J. and Susan J. Gordon, "Recreational Cyclist: The Relationship Between Low Back Pain and Training Characteristics," *International Journal of Exercise Science*, 2010, Vol. 3, No. 3, pp. 79–85.

Sennet, Brian, *Cycling Injuries, American Orthopedic Society for Sports Medicine*, 2009.